An Introduction to Nonsmooth Analysis

An Introduction to Nonsmooth Analysis

Juan Ferrera

ELSEVIER

AMSTERDAM • BOSTON • HEIDELBERG • LONDON
NEW YORK • OXFORD • PARIS • SAN DIEGO
SAN FRANCISCO • SINGAPORE • SYDNEY • TOKYO
Academic Press is an imprint of Elsevier

Notices
Knowledge and best practice in this field are constantly changing. As new research and experience broaden our understanding, changes in research methods, professional practices, or medical treatment may become necessary. Practitioners and researchers must always rely on their own experience and knowledge in evaluating and using any information, methods, compounds, or experiments described herein. In using such information or methods they should be mindful of their own safety and the safety of others, including parties for whom they have a professional responsibility.

To the fullest extent of the law, neither the Publisher nor the authors, contributors, or editors, assume any liability for any injury and/or damage to persons or property as a matter of products liability, negligence or otherwise, or from any use or operation of any methods, products, instructions, or ideas contained in the material herein.

British Library Cataloguing-in-Publication Data
A catalogue record for this book is available from the British Library

Library of Congress Control Number
A catalog record for this book is available from the Library of Congress

ISBN: 978-0-12-800731-0

For information on all Academic Press publications
visit our website at store.elsevier.com

Printed and bound in United States of America

14 15 16 11   10   9 8 7 6 5 4 3 2

This book has been manufactured using Print On Demand technology. Each copy is produced to order and is limited to black ink. The online version of this book will show color figures where appropriate.

to Alex and Asun

CONTENTS

Die Eule der Minerva beginnt erst mit
einbrechenden Dämmerung ihren Flug
G.W.F. Hegel

By studying nonsmooth analysis we attempt to extend differentiability and, more specifically, calculus to a broader setting. It is under this scope that this book must be read, and the reason for which we will be interested in extremes, chain rules, mean value, differential equations and so on throughout the book. We proceed to present the different chapters contained in this book.

The first chapter has an introductory character. We start by presenting lower and upper limits of sequences and functions. Immediately after we introduce lower and upper semicontinuous functions, listing their elementary properties and characterizing continuity through semicontinuity.

One of the most interesting aspects of the theory developed in this book is the double points of view, both geometrical and analytical, which may be employed while introducing concepts. The keystone of this duality is the epigraph (or equivalently, the hypograph). Hence we define these objects as soon as possible, and also characterize both upper and lower semicontinuity in terms of hypographs and epigraphs.

Another important feature that arises in this chapter is the use of extended valued functions. Although one may think that it is more difficult to deal with extended valued functions, the truth is that analysis is easier if we work with these functions. Among other reasons, because we may extend the original functions from their domain to the whole space assigning an infinite value where they were not defined.

The main goal of nonsmooth analysis is to extend differentiable tools to the nonsmooth setting, therefore in this chapter we also introduce a short review of differentiable functions, in order to recall the elementary ideas of differential calculus.

We finish the chapter with two important and non trivial results. They are good examples of the kind of problems, as well as the tools, that nonsmooth analysis deals with.

We first introduce Moreau's envelope, depending on a parameter λ, of a lower semicontinuous function. We prove that Moreau's envelope is Lipschitz, and that it approximates the original function. Secondly, we prove a finite dimensional version of Ekeland's Variational Principle that establishes that lower semicontinuous bounded below functions under small perturbations always attain *a strong minimum*.

Chapter 2 is dedicated to convexity. The natural setting for convexity is a Banach space, and many of the results presented here are true in such a frame. Nevertheless, we develop the theory for Hilbert spaces, presenting also finite dimensional results. Sometimes we prove more general results only for finite dimensional spaces, in order to avoid a deeper study of Hilbert spaces, which is beyond the aim of this book.

We start by introducing elementary properties of convex sets and functions. In particular, the characterization of convex functions via the convexity of their epigraph. We also study the distance function to a convex set.

We then prove that finite dimensional convex functions are continuous. Among the different proofs that can be considered, we have chosen to derive it from the characterization of the interior of the epigraph via the Line Segment Principle, which we establish for Hilbert spaces in general.

We also present Minkowski's Separation Theorem and many of its interesting consequences. We again omit the more general Hahn-Banach Theorem, as it requires more elaborate tools.

Finally, we include in this chapter two less known but nevertheless very interesting results: the properties of the Moreau envelope of a convex function, in the finite dimensional case, and the characterization of twice differentiable convex functions.

Chapter 3 establishes subdifferential calculus for convex functions. Many of the results that we will prove in this chapter are particular cases of results that we will establish in the following chapters. However, in order to motivate the concepts, as well as for historical reasons, we opted for this approach. The subdifferential may be unclear at the beginning, however the idea of the epigraph support is quite natural, and in my opinion the possible redundancies and repetitions that occur later are worth while.

We give a proof of the non emptiness of the subdifferential, which we restrict to the finite dimensional case, but that works in the general case invoking Hahn-Banach's Theorem. The main result of this chapter is the equivalence between the different definitions of the subdifferential, including the proximal one. We finish the chapter with two examples that study the subdifferential of two convex functions: the distance to a convex set, and the maximum function.

Once we have introduced the subdifferential of convex functions, the next step is to generalize this concept to a broader situation. This is the goal of Chapter 4, that together with Chapter 5, constitute the core of this book. The natural setting for the subdifferential is the one of lower semi continuous functions. We start the chapter with a very important theorem that proves the equivalence between the two natural definitions of the subdifferential, namely the traditional one via lower limits, and the viscosity one through differentiable supports. We introduce also the superdifferential for upper semicontinuous functions, and then we prove some of the elementary properties of both the sub and super differential. We will characterize differentiable functions as continuous functions that are both subdifferentiable and superdifferentiable.

Throughout this chapter we study the distance function. This has evident interest in itself due to the importance of that function. We take advantage of this study to introduce other aspects of the subdifferential. In particular, for the finite dimensional setting, we present the subgradients (i.e. the subdifferential elements), as the supports of an auxiliary function: the subderivative. From the geometrical point of view, we introduce the normal and tangent cones, and prove, for the finite dimensional case, the important Theorem that characterizes the subdifferential via the regular normal cone of the epigraph.

The next concept that we introduce is the density of subdifferentiability points. In this general, non convex, setting the subdifferential may be empty at some points, however the set of point where it is non empty is dense. We present the result in the finite dimensional case only.

We finish the chapter with the introduction of the proximal subdifferential, comparing it to Frechet's subdifferential, and proving some of its properties.

Chapter 5 is dedicated to calculus. As we remarked previously, nonsmooth analysis tries to extend differentiable techniques to non differentiable functions. For this reason this is the central chapter. Almost all the results presented here are consequences of the powerful Fuzzy Sum Rule. This result is true in Hilbert spaces in general, however we prove it only for functions defined on $\mathbb{R}^n$. This is due to the fact that for our proof we require compactness of balls, but in the general case this tool has to be replaced by an infinite dimensional Variational Principle, and its proof overflows the scope of this book. The Fuzzy Sum Rule always appears in calculus, either directly or indirectly. Hence we restrict to the finite dimensional framework in the chapter.

As a consequence of this result, we present also a short discussion about constrained minima, and a Fuzzy Chain Rule when the first acting function is C^1. The search of exact rules motivates the introduction of regular functions. Therefore we define them, studying some of their elementary properties. We need to remark that our definition of regular functions is different and not equivalent to the definition of other authors; however under mild conditions these definitions agree.

We finish the chapter with two important calculus results. The first one is the Mean Value Theorem, for which we prove that although a Mean Value Equality is not true even for nice functions, we may use it to present two interesting results, namely the Mean Value Inequality and a sub-super differentiable version of the Theorem. The second one is a deep study of the decreasing character of several variable scalar functions. We present the Multidirectional Mean Value Theorem, for which we only prove a differentiable version. Even though we avoid giving its complete proof, we deduce, among other important consequences of this Theorem, the useful Decrease Principle.

The main subject of Chapter 6 is the generalized gradient of a Lipschitz function. We begin this chapter studying Lipschitz functions, characterizing their regularity, and introducing a new interesting class of Lipschitz functions: the strictly differentiable ones.

We introduce the generalized gradient as the set of supports of the generalized directional derivatives; given that these derivatives depend on the directions in a convex way, the generalized gradient can be established as the subdifferential of the function defined via those generalized derivatives. In other words, we may take advantage of the theory developed previously to study the generalized gradient properties. Among other properties, we characterize the generalized gradient as the convex hull of the set of limits of gradients, assuming without proof Rademacher's Theorem, that is: Lipschitz functions are differentiable a.e.

This last result gives us an idea of how to extend these nonsmooth concepts to vector functions. In this sense we define the Generalized Jacobian, which allows us to set the powerful and general Lipschitz Inverse Function Theorem.

We finish the chapter by introducing the graphical derivative and coderivative. These concepts represent another way of defining nonsmooth extensions of the differential in the vectorial case. They have an advantage with respect to the generalized Jacobian, namely, that instead of Lipschitz's property, we only require continuity of the functions. The graphical derivative appears to be closely related to the one-side directional derivatives. For Lipschitz functions we prove an interesting characterization of the Generalized Jacobian (and the generalized gradient in the scalar case) in terms of the graphical coderivative.

In the last chapter of this book we present some applications. Nonsmooth analysis is a powerful tool since it inherits many of the results of classical differential analysis. Therefore it has many interesting applications in a wide range of fields: from differential equations to mathematical economy passing through operational research or functional analysis. However we will restrict our attention to three of them: flow invariant sets, viscosity solutions of a differential equation, and solving general equations whilst proving, as a consequence, the existence of fixed points.

With respect to the first problem, we characterize flow invariant sets of an autonomous system $x' = \varphi(x)$ in terms of the geometrical behavior on the border of the set of the function φ.

In the second section of this chapter we introduce viscosity solutions of a first order Hamilton-Jacobi equation. We prove the "viscosity" character of viscosity solutions, and present, without proof, existence and uniqueness results.

We finish the chapter and the book with a section devoted to the study of sufficient conditions in order to guarantee that an equation $F(x) = 0$ has a solution, applying the result to finding fixed points, and inverse functions.

A final word about the problems collection that we include. Each chapter incorporates a problems section, with the aim of presenting examples and counterexamples that illustrate the theory, helping to understand it. However we also take advantage of these sections to introduce new concepts and easy results making the core of the book a bit lighter.

Juan Ferrera

ACKNOWLEDGMENT

I wish to thank my colleagues at Universidad Complutense de Madrid who read and commented upon early versions of the manuscript, particularly Professor D. Azagra. I wish to thank also Professor G. Beer of California State University for interesting chats about Lipschitz functions and specially for suggesting some examples. Ph.D. Nadia Smith helped me with the figures and technical support in general.

Juan Ferrera

CHAPTER 1

Basic Concepts and Results

This first chapter has an introductory nature. Upper and lower limits of sequences and functions are important tools in analysis, for that reason we will begin the chapter defining them rigorously. Then we will introduce semicontinuous functions which constitute the most general class of functions for nonsmooth purposes. A glance over differentiability will complete this short review. Throughout the book X will be a real Hilbert space. $|x|$ will denote the norm of a element $x \in X$. B will denote the unit ball. Depending on the students' level, we will sometimes consider that $X = \mathbf{R}^n$, or even that $X = \mathbf{R}$.

1.1 UPPER AND LOWER LIMITS

Let us start with a couple of definitions.

DEFINITION 1.1. For a sequence of real numbers $\{x_n\}$, we define the upper limit, respectively, the lower limit, as

$$\limsup_n x_n = \lim_n \left[\sup\{x_k : k \geq n\} \right]$$

respectively

$$\liminf_n x_n = \lim_n \left[\inf\{x_k : k \geq n\} \right]$$

These limits always exist, and belong to $[-\infty, +\infty]$, since the sequences $\left[\sup\{x_k : k \geq n\} \right]$, and $\left[\inf\{x_k : k \geq n\} \right]$ are decreasing and increasing, respectively. It is not difficult to see that these limits can be characterized in terms of subsequences.

PROPOSITION 1.2. For a real sequence $\{x_n\}$, a value $\alpha \in [-\infty, +\infty]$ satisfies $\alpha = \limsup_n x_n$ if and only if there is a subsequence $\{x_{n_k}\}$ such that $\alpha = \lim_k x_{n_k}$ and for every convergent subsequence $\{x_{n_j}\}$ we have $\lim_j x_{n_j} \leq \alpha$.

PROOF. Let $\alpha = \limsup_n x_n$. If $\alpha = +\infty$ then $\sup\{x_k : k \geq n\} = +\infty$ for every n, in particular $\sup\{x_n\} = +\infty$, which implies that there is a subsequence of $\{x_n\}$ that converges to $+\infty$. If $\alpha = -\infty$, then for every k there is a n_k, which we assume satisfies $n_k < n_{k+1}$, such that

An Introduction to Nonsmooth Analysis. http://dx.doi.org/10.1016/B978-0-12-800731-0.00001-1

$\sup\{x_n : n \geq n_k\} < -k$. in other words, $x_n < -k$ for every $n \geq n_k$ which implies $\lim x_n = -\infty$.

If $\alpha \in \mathbf{R}$ then for every positive integer j, there is a k_j such that $\sup\{x_n : n \geq k\} \in \left(\alpha - \frac{1}{j}, \alpha + \frac{1}{j}\right)$ for every $k \geq k_j$, hence there is an index $n_j \geq k_j$ such that $x_{n_j} \in \left(\alpha - \frac{1}{j}, \alpha + \frac{1}{j}\right)$. It is clear that the sequence $\{x_{n_j}\}$ converges to α.

Let us assume that a subsequence $\{x_{n_k}\}$ converges to a value $r \in [-\infty, +\infty]$. If $r = -\infty$, then $r \leq \limsup_n x_n$ trivially. If $r = +\infty$ then for every positive M there is an index k_0 such that $x_{n_k} \geq M$ for every $k \geq k_0$ therefore $\sup\{x_n : n \geq n_{k_0}\} \geq M$ and consequently $\limsup_n x_n = +\infty$. Finally, if $r \in \mathbf{R}$ we have that $\limsup_n x_n \geq r$ since $\sup\{x_n : n \geq n_k\} \geq x_{n_k}$.

Conversely, let $\alpha = \lim_k x_{n_k}$ for a given subsequence. We have

$$\alpha = \lim_k x_{n_k} \leq \lim_k \left[\sup\{x_n : n \geq n_k\}\right]$$
$$\leq \lim_k \left[\sup\{x_n : n \geq k\}\right] = \limsup_n x_n.$$

The other condition satisfied by α implies that $\alpha \geq \limsup_n x_n$, since we have just proved that there is a subsequence converging to the upper limit. $\qquad\square$

In a similar way we have

PROPOSITION 1.3. *For a real sequence $\{x_n\}$, a value $\alpha \in [-\infty, +\infty]$ satisfies $\alpha = \liminf_n x_n$ if and only if there is a subsequence $\{x_{n_k}\}$ such that $\alpha = \lim_k x_{n_k}$ and for every convergent subsequence $\{x_{n_j}\}$ we have $\lim_j x_{n_j} \geq \alpha$.*

It is clear that $\limsup_n x_n \geq \liminf_n x_n$. When this inequality is an equality, they coincide with the limit.

PROPOSITION 1.4. *For a real sequence $\{x_n\}$, we have that $\{x_n\}$ converges if and only if $\limsup_n x_n = \liminf_n x_n$. In this case $\lim_n x_n = \limsup_n x_n = \liminf_n x_n$.*

The proof of this proposition is left to the reader as an exercise.

We are going to introduce the concept of upper and lower limits of a real function, $f : X \to \mathbf{R}$.

DEFINITION 1.5.   For a function $f : X \to \mathbf{R}$, we define the upper limit at $x_0 \in X$ as

$$\limsup_{x \to x_0} f(x) = \lim_{\delta \searrow 0} \left[\sup\{ f(x) : x \in B(x_0, \delta) \} \right]$$

The lower limit at x_0 is defined similarly as

$$\liminf_{x \to x_0} f(x) = \lim_{\delta \searrow 0} \left[\inf\{ f(x) : x \in B(x_0, \delta) \} \right]$$

These limits always exist (possibly taking values $\pm\infty$) and we have

$$\limsup_{x \to x_0} f(x) \geq \liminf_{x \to x_0} f(x)$$

It is clear that instead of limits for $\delta \searrow 0$ we can take limits for n with the sup (or inf) over $B\left(x_0, \frac{1}{n} \right)$. The proofs of the following results are standard and we will omit them.

PROPOSITION 1.6.   *The upper (respectively lower) limit of a function $f : X \to \mathbf{R}$ at a point x_0, α, is characterized by the following properties: there exists a sequence $\{x_n\}$ converging to x_0 such that $\lim_n f(x_n) = \alpha$, and for every sequence $\{x_n\}$ converging to x_0 that satisfies that $\{f(x_n)\}$ converges, we have $\lim_n f(x_n) \leq \alpha$ (respectively $\lim_n f(x_n) \geq \alpha$).*

Sometimes we omit the point x_0 while taking sup or inf on $B(x_0, \delta)$. We will specify this by writing

$$\limsup_{x \to x_0, x \neq x_0} f(x) \text{ or } \liminf_{x \to x_0, x \neq x_0} f(x)$$

These limits give us the following proposition.

PROPOSITION 1.7.   *For a function $f : X \to \mathbf{R}$ and a point $x_0 \in X$, we have that $\lim_{x \to x_0} f(x)$ exists if and only if*

$$\limsup_{x \to x_0, x \neq x_0} f(x) = \liminf_{x \to x_0, x \neq x_0} f(x)$$

In this case the three limits agree.

1.2 SEMICONTINUITY

DEFINITION 1.8.   We start by recalling a well-known definition. We say that a real function $f : X \to \mathbf{R}$ is continuous at $x \in X$ provided that for every $\varepsilon > 0$ there is a $\delta > 0$ such that $|f(x) - f(y)| < \varepsilon$ whenever $|x - y| < \delta$. If this property holds for every $x \in X$, we say that f is continuous on X.

It is easy to see that a function f is continuous at x if and only if for every sequence $\{x_n\}$ converging to x, $\{f(x_n)\}$ converges to $f(x)$.

Continuous functions have many good properties; they attain maxima and minima over compacts for instance. Many of these properties hold under weaker continuity conditions. The aim of this section is to introduce such conditions.

DEFINITION 1.9. We say that a real function $f : X \to \mathbf{R}$ is lower semicontinuous (lsc), respectively upper semicontinuous (usc), at x_0, provided that $f(x_0) \leq \liminf f(x_n)$, respectively $f(x_0) \geq \limsup f(x_n)$, for every sequence $\{x_n\} \subset X$ satisfying $\lim_n x_n = x_0$. If the property holds for every point $x_0 \in X$ we say that f is lsc, usc respectively, on X.

It is not difficult to see that lower semicontinuity at x_0 is equivalent to

$$f(x_0) \leq \liminf_{x \to x_0} f(x)$$

while upper semicontinuity is characterized by

$$f(x_0) \geq \limsup_{x \to x_0} f(x)$$

The following proposition gives us a ε-δ characterization of lower semicontinuity.

PROPOSITION 1.10. f is lsc at x_0 if and only if for every $\varepsilon > 0$ there is a $\delta > 0$ such that $f(x_0) - f(x) < \varepsilon$ whenever $|x - x_0| < \delta$. A similar result holds for usc functions

PROOF. Assume that f is lsc at x_0. Let $\alpha = \liminf_{x \to x_0} f(x)$ then for every $\varepsilon > 0$ there is a $\delta > 0$ such that $|\alpha - \inf\{f(x) : x \in B(x_0, \delta)\}| < \varepsilon$. For every $x \in B(x_0, \delta)$ we have

$$f(x_0) - f(x) \leq \alpha - f(x) \leq \alpha - \inf\{f(x) : x \in B(x_0, \delta)\} < \varepsilon$$

Conversely, let $\varepsilon > 0$, choose a positive δ_0 such that $f(x_0) - f(x) < \frac{\varepsilon}{2}$ for every $x \in B(x_0, \delta_0)$. We take a point $\tilde{x} \in B(x_0, \delta_0)$ such that

$$f(\tilde{x}) < \inf\{f(x) : x \in B(x_0, \delta_0)\} + \frac{\varepsilon}{2}$$

we have

$$f(x_0) - \inf\{f(x) : x \in B(x_0, \delta)\} < f(x_0) - f(\tilde{x}) + \frac{\varepsilon}{2} < \varepsilon$$

for every positive $\delta \leq \delta_0$. Let us observe also that the left-hand side of the inequality is positive, hence $|f(x_0) - \inf\{f(x) : x \in B(x_0, \delta)\}| < \varepsilon$ for every positive $\delta \leq \delta_0$. We have proved that

$$f(x_0) = \liminf_{x \to x_0} f(x) \qquad \qquad \square$$

The following question arises: have we proved more than lower semi-continuity? Of course not, since we have defined lower limits in such a way that the inequality $\liminf_{x \to x_0} f(x) \leq f(x_0)$ always holds. The following result is a mere exercise.

PROPOSITION 1.11.   *f is continuous if and only if it is both lsc and usc.*

The following proposition summarizes some elementary properties of semicontinuous functions. We suggest the reader work through the details of the proof.

PROPOSITION 1.12.   *Let $f, g : X \to \mathbf{R}, \lambda > 0$. We have*

(i)   f is lsc if and only if $-f$ is usc
(ii)   If f, g are lsc, respectively usc, then $f + g$ is lsc, respectively usc.
(iii) If f is lsc, respectively usc, then λf is also lsc, respectively usc.

One of the goals of this book is to establish the involved ideas in a geometrical form as well as an analytical one. For this we introduce two sets associated to a given function, which will allow us to present the analytical properties that we have just defined as geometric properties.

DEFINITION 1.13.   For a given function $f : X \to \mathbf{R}$, we define its epigraph, respectively hypograph, as the following set: $epif = \{(x, r) \in X \times \mathbf{R} : f(x) \leq r\}$, respectively $hypof = \{(x, r) \in X \times \mathbf{R} : f(x) \geq r\}$.

Despite its simplicity the next result remarks the relation between analytic and geometrical properties.

PROPOSITION 1.14.   *Let $f : X \to \mathbf{R}$ be a function, we have that f is lsc if and only if $epif$ is a closed set, similarly f is usc if and only if $hypof$ is closed (see Fig. 1.1).*

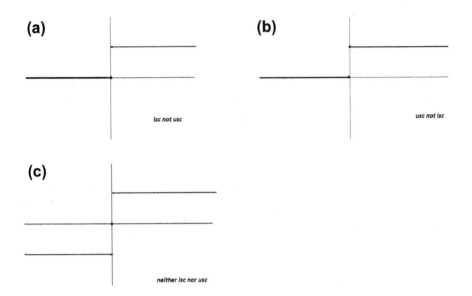

(a)

lsc not usc

(b)

usc not lsc

(c)

neither lsc nor usc

Fig. 1.1 Graphs of semicontinuous functions.

PROOF.   Assume that f is lsc, let $\{(x_n, r_n)\} \subset epif$ be a sequence converging to (x, r), $f(x_n) \leq r_n$ implies that $\liminf f(x_n) \leq r$, hence $f(x) \leq r$ since f is lsc, and consequently $(x, r) \in epif$.

Conversely we assume that $epif$ is closed. Fix a point x_0, given a sequence $\{x_n\}$ converging to x_0, such that $\lim_n f(x_n) = \liminf_{x \to x_0} f(x)$, we have that $(x_n, f(x_n)) \in epif$ and consequently $(x, \liminf_{x \to x_0} f(x))$ belongs to $epif$ too since it is closed. This implies that $f(x) \leq \liminf_{x \to x_0} f(x)$, in other words: f is lsc at x_0. The proof for usc is similar.   □

As a consequence of the preceding results, a continuous function satisfies that its graph, which can be represented as $epif \cap hypof$, is closed. The converse is not true, consider for instance the function $f : \mathbf{R} \to \mathbf{R}$ defined as $f(x) = \frac{1}{x}$ if $x > 0$, $f(x) = 0$ otherwise. Its graph and epigraph are closed, but $hypof$ does not have this property.

An easy and important consequence of the preceding proposition is the stability of semicontinuous functions under some operations. For instance:

COROLLARY 1.15.   *Let* $\{f_i\}_{i \in I}$ *be a family of lsc, respectively usc functions. Then* $\sup_{i \in I} f_i$ *is lsc, respectively* $\inf_{i \in I} f_i$ *is usc.*

Fig. 1.2 Continuity is not preserved by sup.

PROOF. The result is a consequence of the fact that

$$epi\,(\sup_{i \in I} f_i) = \bigcap_{i \in I} epi\, f_i \quad \text{and} \quad hypo(\inf_{i \in I} f_i) = \bigcap_{i \in I} hypo\, f_i \qquad \square$$

These results are no longer true for continuous functions. The sup of the following family of continuous functions is not continuous.

Example. $f_n : \mathbf{R} \to \mathbf{R}$ defined by 0 if $x \leq 0$, 1 if $x \geq \frac{1}{n}$, and nx if $x \in \left[0, \frac{1}{n}\right]$ (see Fig. 1.2).

A well-known result establishes that every continuous function attains its minimum in any compact set. The next result extends this property.

PROPOSITION 1.16. *Let K be a compact subset of X. Every lsc function $f : X \to \mathbf{R}$ attains its minimum with respect to K.*

PROOF. Let $r_0 = \inf\{f(x) : x \in K\}$. If $r_0 = -\infty$, there would be a sequence $\{x_n\} \subset K$ such that $\lim f(x_n) = -\infty$. We may assume without loss of generality that $\{x_n\}$ converges to $x_0 \in K$ by compactness. Lower semicontinuity of f would imply that $f(x_0) \leq -\infty$, which is not possible. Hence $r_0 \in \mathbf{R}$, and we take a sequence $\{x_n\} \subset K$, that we may consider convergent to a $x_0 \in K$ again, such that $\lim f(x_n) = r_0$. Finally, $f(x_0) \leq \lim f(x_n) = r_0$ implies $f(x_0) = r_0$, hence x_0 is a minimum. $\square$

Sometimes we will consider functions defined on a subset instead of on the whole space X. The set where the function is defined is called domain of f, and we will denote it by *domf*. In this case, continuity concepts are defined as above, imposing that all the arguments lie in *domf*.

We will sometimes allow our functions to reach the value $+\infty$, in other words: $f : X \to (-\infty, +\infty]$. Many definitions make sense in this new context, since we can calculate limits and do algebraic operations (except for $-(+\infty)$!). Accepting this new value, we may extend functions defined in a subset of X to the whole space, assuming that the value of the function is $+\infty$ out of its domain. Some properties of the function are stable under this kind of extension, the minimum value for instance. The reason for choosing the value $+\infty$ instead of $-\infty$ is precisely because we are going to deal with minima, and *lsc* functions. Replacing function f by $-f$, minima are maxima, lower semicontinuity transforms into upper semicontinuity, and we extend the function to $-\infty$. Both derivations are trivially equivalent. However we will not admit the function $f(x) \equiv +\infty$. Even though it is probably clear how to define *lsc* functions for functions taking values in $(-\infty, +\infty]$, we still write it.

DEFINITION 1.17. We say that a function $f : X \to (-\infty, +\infty]$ is *lsc* at a point $x_0 \in X$ if $f(x_0) \leq \liminf_{x \to x_0} f(x)$.

Let us observe that lower limits are defined exactly as in Definition 1.5 if the range of the function is $(-\infty, +\infty]$. If $x_0 \notin dom f$, then *lsc* implies that $\liminf_{x \to x_0} f(x)$, and consequently $\lim_{x \to x_0} f(x)$ are $+\infty$. Upper semicontinuity is defined for functions $f : X \to [-\infty, +\infty)$ in a similar way.

An extremely useful function in this setting is the *indicator* function δ_A of a set A, defined by

$$\delta_A(x) = 0 \text{ if } x \in A, \qquad \delta_A(x) = +\infty \text{ if } x \notin A$$

It is left as an exercise for the reader to see that δ_A is *lsc* if and only if A is closed.

1.3 DIFFERENTIABILITY

A well-known concept is differentiability. However we refresh some ideas.

DEFINITION 1.18. We say that a real function, $f : X \to \mathbf{R}$ is differentiable at $x \in X$ provided that there is a linear function $L : X \to \mathbf{R}$ such that

$$\lim_{h \to 0} \frac{f(x+h) - f(x) - L(h)}{|h|} = 0$$

we denote this linear function L by $df(x)$ or by $f'(x)$, and we say that it is the differential of f at x.

This property is equivalent to the existence of a vector $\nabla f(x) \in X$ such that

$$\lim_{h \to 0} \frac{f(x+h) - f(x) - \langle \nabla f(x), h \rangle}{|h|} = 0$$

When this property arises at every $x \in X$, we say that f is differentiable. It is clear that differentiability implies continuity. On the other hand for every $v \in X$, if we particularize $h = tv$ with $t \to 0$ in the preceding equality, we obtain

$$\langle \nabla f(x), v \rangle = \lim_{t \to 0} \frac{f(x + tv) - f(x)}{t}$$

we say that this limit is the directional derivative of f at x in the direction v, and we denote it by $f'_v(x)$ or by $d_v f(x)$. The relationship between all these notations is summarized in the following list of equalities:

$$df(x)(v) = d_v f(x) = f'_v(x) = f'(x)(v) = \langle \nabla f(x), v \rangle.$$

From the preceding derivation it is clear that a differentiable function has all its directional derivatives. The converse is not true. The *lsc* function $f : \mathbf{R}^2 \to \mathbf{R}$ defined by $f(x, x^2) = 0$ and $f(x, y) = 1$ if $y \neq x^2$ is not continuous at 0, nor differentiable of course, however $f'_v(0, 0) = 0$ for every v.

Directional derivatives of a differentiable function depend linearly on the directions. This property provides us with a simple method to construct functions with directional derivatives that lack differentiability, for instance: $f : \mathbf{R}^2 \to \mathbf{R}$ defined as $f(x, y) = x$ if $y = 0$ and $f(x, y) = 0$ otherwise, cannot be differentiable since $d_{(1,0)} f(0, 0) = 1$, $d_{(-1,0)} f(0, 0) = -1$, but $d_{(v_1, v_2)} f(0, 0) = 0$ in all the other cases. However, let us observe that directional derivatives always have the following property:

PROPOSITION 1.19. *If $d_v f(x)$ exists, then $d_{\lambda v} f(x) = \lambda d_v f(x)$ for every $\lambda \in \mathbf{R}$.*

As we have seen in a preceding example, a nondifferentiable function may have all its directional derivatives, and moreover the derivatives may depend linearly on the directions. This last property is presented in the next definition.

DEFINITION 1.20. We say that a function $f : X \to \mathbf{R}$ is Gâteaux differentiable at $x \in X$, if there is an element of X, denoted by $f'_G(x)$, such that $d_v f(x) = \langle f'_G(x), v \rangle$ for every $v \in X$. We say that $f'_G(x)$ is the Gâteaux derivative of f at x.

In order to avoid confusion between Gâteaux differentiability and differentiability, differentiable functions are usually called Frechet differentiable functions.

It is an exercise to observe that if $X = \mathbf{R}$, then Gâteaux and Frechet differentiability agree. On the other hand there is only one significative directional derivative, namely $d_1 f(x) = f'_1(x)$ that obviously coincides with $f'(x)$.

In the same way as when introducing continuity ideas, the functions do not need to be defined in the whole space. However, while working with differentiability concepts, we will assume that $dom f$ is open.

1.4 TWO IMPORTANT THEOREMS

We introduce a new concept now: the Lipschitz property. In our framework, this property is sort of a middle way between continuity and differentiability; and Lipschitz functions, sometimes, play the role of differentiable functions in nonsmooth analysis. Although we will study Lipschitz functions in depth in Chapter 6, we define them now.

DEFINITION 1.21. Let $A \subset X$, we say that a real function, $f : X \to \mathbf{R}$ is Lipschitz on A, provided that there is a positive constant K, such that $|f(x) - f(y)| \le K |x - y|$ for every $x, y \in A$. A locally Lipschitz function is a function that is Lipschitz in a neighborhood of every point of X.

If we want to specify the constant of the definition, we say that f is K-Lipschitz. It is clear that if a function is K-Lipschitz, then it is K'-Lipschitz for every $K' \ge K$.

Convolution is a well-known method to regularize and approximate functions in the differentiable setting. In our context, the inf-convolution plays this role.

DEFINITION 1.22. For a *lsc* function $f : X \to (-\infty, +\infty]$, bounded below, and a positive parameter λ, we define the Moreau envelope f_λ as:

$$f_\lambda(x) = \inf_{w \in X} \left\{ f(w) + \frac{1}{2\lambda}|w - x|^2 \right\}$$

The function f_λ is clearly well defined, finite and $b \le f_\lambda(x) \le f(x)$ provided that b is a lower bound for f. But f_λ has other nice properties.

THEOREM 1.23. *If $f : X \to (-\infty, +\infty]$ is a lsc function, bounded below, and $A \subset X$ is a bounded set, then f_λ is Lipschitz on A.*

PROOF. Let $x_0 \in domf$, we have that $f_\lambda(x) \le f(x_0) + \frac{1}{2\lambda}|x - x_0|^2$, and consequently $M = \sup\{f_\lambda(x) : x \in A\}$ is finite.

Now let $x, y \in A$ and $\varepsilon > 0$, there is a $z \in X$ such that $f_\lambda(y) + \varepsilon \ge f(z) + \frac{1}{2\lambda}|z - y|^2$. Thus we have:

$$f_\lambda(x) - f_\lambda(y) \le f_\lambda(x) - f(z) - \frac{1}{2\lambda}|z - y|^2 + \varepsilon$$

$$\le f(z) + \frac{1}{2\lambda}|z - x|^2 - f(z) - \frac{1}{2\lambda}|y - z|^2 + \varepsilon$$

$$= \frac{1}{2\lambda}|z - x|^2 - \frac{1}{2\lambda}|y - z|^2 + \varepsilon$$

$$= \frac{1}{2\lambda}|x - y|^2 - \frac{1}{\lambda}\langle x - y, z - y \rangle + \varepsilon$$

$$\le \frac{1}{2\lambda}|x - y|^2 + \frac{1}{\lambda}|x - y||z - y| + \varepsilon$$

$$= \frac{1}{2\lambda}|x - y|(|x - y| + 2|z - y|) + \varepsilon \le K|x - y| + \varepsilon$$

The last inequality holds since $x, y \in A$, which is bounded, and z lies in a bounded set also because $|z - y|^2 \le 2\lambda(f_\lambda(y) + \varepsilon - f(z)) \le 2\lambda(M + \varepsilon - \inf f)$ and consequently $z \in A + \alpha B$ where $\alpha^2 = 2\lambda(M + \varepsilon - \inf f)$.

Reversing the roles of x and y we get $|f_\lambda(x) - f_\lambda(y)| \le K|x - y| + \varepsilon$, and letting $\varepsilon \downarrow 0$, we have that f_λ is Lipschitz on A. □

Once we have proved that Moreau envelopes regularize *lsc* functions, we are going to show that they approximate the original function.

THEOREM 1.24. *Let $f : X \to (-\infty, +\infty]$ be a lsc function bounded below, then $\lim_{\lambda \downarrow 0} f_\lambda(x) = f(x)$ for every $x \in X$.*

PROOF. Let $b \leq f(x)$ for every $x \in X$. Assume first that $f(x) < +\infty$, we have that $f(w) + \frac{1}{2\lambda}|w - x|^2 \geq b + \frac{1}{2\lambda}|w - x|^2 > f(x)$ whenever $|w - x|^2 > 2\lambda(f(x) - b)$ or equivalently if $w \notin \overline{B}\left(x, (2\lambda(f(x) - b))^{\frac{1}{2}}\right)$. We denote $(2\lambda(f(x) - b))^{\frac{1}{2}} = r_\lambda$, which goes to 0 when $\lambda \downarrow 0$. Hence

$$f_\lambda(x) = \inf_{w \in \overline{B}(x, r_\lambda)} \left(f(w) + \frac{1}{2\lambda}|w - x|^2 \right)$$

since $f_\lambda(x) \leq f(x)$. We deduce from the formula above that

$$f_\lambda(x) \geq \inf_{w \in \overline{B}(x, r_\lambda)} (f(w))$$

and consequently

$$f(x) \geq \limsup_{\lambda \downarrow 0} f_\lambda(x) \geq \liminf_{\lambda \downarrow 0} f_\lambda(x) \geq \lim_{\lambda \downarrow 0} \inf_{w \in \overline{B}(x, r_\lambda)} (f(w))$$
$$= \liminf_{w \to x} f(w) \geq f(x)$$

by lower semicontinuity. Hence $\lim_{\lambda \downarrow 0} f_\lambda(x) = f(x)$. The other case is trivial since $\lim_{w \to x} f(w) = \liminf_{w \to x} f(w) = +\infty$ whenever $f(x) = +\infty$. □

The distance function to a set S is defined by $d_S(x) = \inf_{y \in S} |y - x|$. The most significant situation is when S is closed, since in this case $d_S(x) = 0$ implies $x \in S$. It is easy to see that $d_S^2 = (\delta_S)_{\frac{1}{2}}$, and considering that δ_S is lsc and bounded below, we have that d_S^2 is Lipschitz and consequently d_S is continuous. It is not difficult to directly prove a stronger result.

PROPOSITION 1.25.   *The distance function d_S is Lipschitz for every $S \subset X$*

PROOF. Let $x, y \in X$, and $\varepsilon > 0$. There is a $z \in S$ such that $d_S(x) + \varepsilon > |x - z|$. We deduce

$$d_S(y) - d_S(x) < d_S(y) - |x - z| + \varepsilon \leq |y - z| - |x - z| + \varepsilon \leq |y - x| + \varepsilon$$

letting $\varepsilon \downarrow 0$ we get $d_S(y) - d_S(x) \leq |x - y|$. Changing the roles of x and y we have

$$|d_S(y) - d_S(x)| \leq |x - y|.$$

Hence d_S is 1-Lipschitz. □

The last theorem that we present in this chapter is a good example of a nonsmooth result. We will prove the case $X = \mathbf{R}^n$, but the result is true in a more general setting. We will not use this result later on, so the reader may skip its proof.

DEFINITION 1.26. We say that a point x_0 strongly minimizes a function f, if $f(x_0) = \inf f$, and moreover every sequence $\{x_n\}$ that satisfies $\lim f(x_n) = \inf f$, converges to x_0 necessarily.

THEOREM 1.27. *(Ekeland's Variational Principle) Let $f : \mathbf{R}^n \to (-\infty, +\infty]$ be a lsc function, bounded below, let $\varepsilon > 0$, and select a x_0 such that $f(x_0) < \inf f + \varepsilon$. Then for every $\delta > 0$ there exists a point $\hat{x}$ such that $|\hat{x} - x_0| < \frac{\varepsilon}{\delta}$, $f(\hat{x}) \leq f(x_0)$, and $\hat{x}$ strongly minimizes the function*

$$\hat{f}(x) = f(x) + \delta|x - \hat{x}|$$

PROOF. The function $f_0(x) = f(x) + \delta|x - x_0|$ is clearly bounded below and *lsc*. If $x \notin B\left(x_0, \frac{\varepsilon}{\delta}\right)$, we have that $f_0(x) > f(x) + \varepsilon \geq \inf f + \varepsilon > f(x_0) = f_0(x_0)$, hence

$$\inf f_0 = \inf_{x \in \overline{B}\left(x_0, \frac{\varepsilon}{\delta}\right)} f_0(x)$$

The *lsc* function f_0 attains its minima on the compact $\overline{B}\left(x_0, \frac{\varepsilon}{\delta}\right)$, this proves that f_0 attains necessarily a minimum. We denote the set of such minima by A, considering that a set of minima of a *lsc* function is always closed, we have that A is a nonempty compact.

We define now a new *lsc* function $\tilde{f} = f + \delta_A$ which is bounded below and consequently attains its minima on the compact A. Let $\hat{x}$ be a minimum of $\tilde{f}$, it also minimizes f_0, that is

$$f(\hat{x}) + \delta|\hat{x} - x_0| \leq f(x) + \delta|x - x_0|$$

for every x. Hence, if we define $\hat{f}(x) = f(x) + \delta|x - \hat{x}|$, we have

$$\hat{f}(\hat{x}) = f(\hat{x}) \leq f(x) + \delta|x - x_0| - \delta|\hat{x} - x_0| \leq f(x) + \delta|x - \hat{x}| = \hat{f}(x)$$

this implies that $\hat{x}$ is a minimum of $\hat{f}$. Let us observe that $|\hat{x} - x_0| < \frac{\varepsilon}{\delta}$ since $\hat{x} \in A \subset \overline{B}\left(x_0, \frac{\varepsilon}{\delta}\right)$, and $f(\hat{x}) = f_0(\hat{x}) \leq f_0(x_0) = f(x_0)$ since $\hat{x}$ minimizes f_0. Let us suppose that there is another minimum $\hat{y}$ of $\hat{f}$.

We would have that $f(\hat{x}) = \hat{f}(\hat{x}) = \hat{f}(\hat{y}) = f(\hat{y}) + \delta|\hat{y} - \hat{x}|$, in particular $f(\hat{y}) < f(\hat{x})$ which implies that $\hat{y} \notin A$, since $\hat{x}$ minimizes f when restricted to A. The fact that $\hat{y}$ does not minimize f_0 implies

$$f(\hat{y}) + \delta|\hat{y} - x_0| > f(\hat{x}) + \delta|\hat{x} - x_0| = f(\hat{y}) + \delta|\hat{y} - \hat{x}| + \delta|\hat{x} - x_0|$$

Then $|\hat{y} - x_0| > |\hat{y} - \hat{x}| + |\hat{x} - x_0|$ which is impossible.

Once we have established that $\hat{x}$ is the only minimum of $\hat{f}$, we have to see that $\hat{f}$ attains that minimum strongly. Let $\{x_n\}$ be a minimizing sequence of $\hat{f}$, that is $\hat{f}(\hat{x}) = \lim \hat{f}(x_n)$, but

$$\lim \hat{f}(x_n) = \lim(f(x_n) + \delta|x_n - \hat{x}|)$$
$$= \lim \sup(f(x_n) + \delta|x_n - \hat{x}|) \geq \lim \sup(\inf f + \delta|x_n - \hat{x}|)$$

hence the sequence $\{x_n\}$ is bounded, converges by compactness, and converges to a minimum since $\hat{f}$ is *lsc*. We conclude that it converges to $\hat{x}$. $\square$

1.5 PROBLEMS

(1) Calculate the upper and lower limits of the following sequences $\{x_n\}$:
$$x_n = (-1)^{3n+1} n^{[(-1)^n]}, \quad x_n = (-1)^n \cos n,$$
$$x_n = (-1)^{2n+1} \left(\frac{n+1}{n}\right)^{[(-1)^n]}.$$

(2) Calculate all the limit points of the sequences in problem 1.
(3) Prove Proposition 1.4.
(4) Prove Proposition 1.6.
(5) Prove Proposition 1.7.
(6) Prove that a function f is *lsc* at a point x_0 if and only if
$$f(x_0) \leq \liminf_{x \to x_0, x \neq x_0} f(x).$$

(7) Give an example of a function for which
$$int(epif) \neq \{(x, r) : f(x) < r\}.$$

(8) Under what conditions do we have
$$int(epif) = \{(x, r) : f(x) < r\}?$$

(9) Prove that a bounded function with closed graph is necessarily continuous.

(10) Prove that a function $f : domf \to \mathbf{R}$ is *lsc* provided that the extended function $\tilde{f} : X \to (-\infty, +\infty]$ defined by $\tilde{f}(x) = f(x)$ if $x \in domf$ and $\tilde{f}(x) = +\infty$ otherwise, is *lsc* too.

(11) Prove that if $domf$ is closed, then the converse is also true.

(12) Prove that $f : X \to (-\infty, +\infty]$ is *lsc* if and only if $epif$ is closed

(13) Prove that the set of minima of a *lsc* function is closed.

(14) Draw the Moreau Envelope f_λ, for an arbitrary $\lambda > 0$, of the function $f : \mathbf{R} \to \mathbf{R}$ defined by $f(t) = 1$ if $t > 0$ and $f(t) = 0$ otherwise.

(15) Draw the Moreau Envelope f_λ, for an arbitrary $\lambda > 0$, of the function $\delta_{[-1,1]}$.

CHAPTER 2

Convex Functions

In this chapter we introduce convex sets and functions, prove some properties of convex functions, in particular their continuity in the finite dimensional setting. We also study the distance function to a convex set, and the Moreau envelope of a convex function. We also prove the important Minkowski Separation Theorem, as well as some interesting Corollaries. We end the chapter by characterizing differentiable convex functions.

2.1 CONVEX SETS AND CONVEX FUNCTIONS

A C in a Hilbert space X is a set with the following property: for every $x, y \in C$, C contains the segment $[x, y] = \{tx + (1 - t)y : t \in [0, 1]\}$ (see for instance, Fig. 2.1). The next proposition summarizes some elementary properties of convex sets:

PROPOSITION 2.1.

 (i) If C is convex, then $\overline{C}$ is also convex.
 (ii) $\bigcap_{i \in I} C_i$ is convex provided that C_i is convex for every $i \in I$.
(iii) C convex and $\lambda \in \mathbf{R}$ implies λC convex.
(iv) If C_1 and C_2 are convex, then $C_1 + C_2$ is also convex.
 (v) If C is convex, then $\lambda_1 C + \lambda_2 C = (\lambda_1 + \lambda_2)C$ for every $\lambda_1, \lambda_2 \geq 0$. Consequently $\lambda_1 C + \lambda_2 C$ is also convex.

Part (v) is the only one that has some difficulty, but to prove it, it is enough to observe that we may assume, without loss of generality, that $\lambda_1 + \lambda_2 = 1$. Part (ii) allows us to define the convex hull of an arbitrary set since the intersection of all the convex sets that contain a given set is itself convex. Let us see it.

DEFINITION 2.2. For a given set $A \subset X$, we define the convex hull of A, coA, as the intersection of all the convex sets that contain A.

Let us observe that coA is well defined since there always is, at least, one convex set that contains A, namely the whole of X. Of course coA is

An Introduction to Nonsmooth Analysis. http://dx.doi.org/10.1016/B978-0-12-800731-0.00002-3

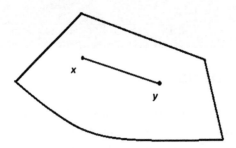

Fig. 2.1 A convex set.

convex, contains A, and as a matter of fact is the smallest convex set that contains A. The next proposition characterizes convex hulls.

PROPOSITION 2.3. *Let $A \subset X$, we have*

$$coA = \left\{ \sum_{i=1}^{n} \lambda_i x_i : x_i \in A, \lambda_i \geq 0, \sum_{i=1}^{n} \lambda_i = 1 \right\}. \tag{2.1}$$

PROOF. Let us denote the set $\left\{ \sum_{i=1}^{n} \lambda_i x_i : x_i \in A, \lambda_i \geq 0, \sum_{i=1}^{n} \lambda_i = 1 \right\}$ by D, it is clear that D is convex since

$$t \sum_{i=1}^{n} \lambda_i x_i + (1 - t) \sum_{i=1}^{m} \mu_i y_i \in D$$

for every $t \in [0, 1]$ provided that $\sum_{i=1}^{n} \lambda_i x_i$ and $\sum_{i=1}^{m} \mu_i y_i$ belong to D. It is also clear that $A \subset D$, hence $coA \subset D$.

On the other hand, in order to prove the opposite inclusion, it is enough to observe that every convex set C that contains A contains D too. This is true since the elements of D are convex combinations of elements of $A \subset C$ and consequently belong to C. □

Let us observe that the convex combinations that appear in the characterization of coA have an arbitrary long number of points x_i. However a celebrated theorem due to Caratheodory asserts that it is enough to take $n+1$ points to determine the convex hull of sets in R^n. Convex hulls satisfy the following identity (the proof is left to the reader)

$$co(A + B) = coA + coB \tag{2.2}$$

Another important set while dealing with convexity is the closed convex hull.

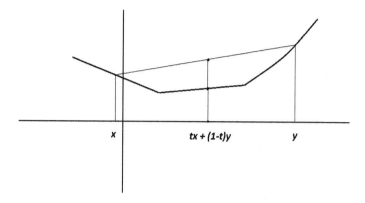

x $tx + (1-t)y$ y

Fig. 2.2 A convex function.

DEFINITION 2.4. Let $A \subset X$, we define the closed convex hull of A, denoted by $\overline{co}A$ as the closure of the convex hull, that is $\overline{co}A = \overline{coA}$.

We will see in the problems section that convex hulls satisfy the following inclusion

$$\overline{co}(A) + \overline{co}(B) \subset \overline{co}(A + B)$$

The equality holds if one of the sets is compact.

We are now going to introduce the concept of a convex function. A convex function is usually defined as a real function, with a convex domain C, that satisfies the following property: $f(tx+(1-t)y) \leq tf(x) + (1-t)f(y)$ for every $x, y \in C$ and every $t \in [0, 1]$. We now introduce convexity for extended valued functions.

DEFINITION 2.5. We say that the function $f : X \to (-\infty, +\infty]$ is convex if $f(tx + (1-t)y) \leq tf(x) + (1-t)f(y)$ holds for every $x, y \in X$ and every $t \in [0, 1]$ (see Fig. 2.2).

Let us observe that for a convex function f, $dom f$ is necessarily convex since otherwise we would have two points x, y such that $f(x), f(y) \in \mathbf{R}$ and a point $z \in [x, y]$ with $f(z) = +\infty$, which is not possible.

It is an exercise to prove that the norm $|.|$ is a convex function. On the other hand the indicator function δ_C is convex if and only if C is convex.

Geometrically, convexity of a function means that the line segment that joins $(x, f(x))$ and $(y, f(y))$ is above the function's graph. In other words: *epi f* is convex. Let us prove it.

PROPOSITION 2.6.   *A function $f : X \to (-\infty, +\infty]$ is convex if and only if its epigraph is convex.*

PROOF.   Let us assume that f is convex. For two points (x_1, r_1), (x_2, r_2) $\in epif$, $x_1, x_2 \in domf$, an arbitrary convex combination, $t(x_1, r_1)+(1-t)$ (x_2, r_2) with $t \in [0, 1]$, necessarily belongs to $epif$ too, because

$$f(tx_1 + (1 - t)x_2) \le tf(x_1) + (1 - t)f(x_2) \le tr_1 + (1 - t)r_2$$

This proves the convexity of $epif$.

Conversely, we have to check that $f(tx+(1-t)y) \le tf(x)+(1-t)f(y)$, where $x, y \in X$ and $t \in [0, 1]$. We may assume without loss of generality that $x, y \in domf$ since otherwise the inequality is trivial. In this case, the inequality holds provided that $(tx +(1-t)y, tf(x)+(1-t)f(y)) \in epif$. And this is true since $(tx + (1 - t)y, tf(x) + (1 - t)f(y)) = t(x, f(x)) + (1 - t)(y, f(y))$, and $(x, f(x))$ as well as $(y, f(y))$ belong to $epif$.   □

Given a function $f : A \to \mathbf{R}$, with $A \subset X$, we may define its epigraph as $epif = \{(x, r) \in A \times \mathbf{R} : f(x) \le r\}$. If we extend the function by means of $\tilde{f}(x) = +\infty$ if $x \notin A$, then we have that $epif = epi\,\tilde{f}$. Therefore f is convex if and only if $\tilde{f}$ is convex. This is another reason that justifies the introduction of the value $+\infty$, as well as our preference in looking for minima instead of maxima!

A consequence of Proposition 2.6 is that the supremum of a family of convex functions is also convex.

COROLLARY 2.7.   *Let $f_i : X \to (-\infty, +\infty]$, $i \in I$, be a family of convex functions. Then $\sup_{i \in I} f_i$ is also convex provided that it is not identically $+\infty$.*

PROOF.   The result is a consequence of the fact that $epi(\sup_{i \in I} f_i) = \bigcap_{i \in I} epif_i$, and that the intersection of a family of convex sets is convex.   □

Another interesting consequence of Proposition 2.6 is that under mild assumptions on S, the distance function d_S is convex. In order to study this function we start by representing its epigraph. The key for this is the following general Lemma that we will use again later.

LEMMA 2.8.   *Let $f, g : X \to (-\infty, +\infty]$ be two functions. We define*

$$h(x) = \inf_{w \in X} (f(w) + g(x - w))$$

Then $epi\, h = epi\, f + epi\, g$ *if the* inf *in the definition of* h *is attained always.*

PROOF. If $(x, r) \in epi\, f$ and $(y, s) \in epi\, g$, we have $r + s \geq f(x) + g(y) = f(x) + g(y + x - x) \geq h(y + x)$, hence $(x + y, r + s) \in epi\, h$. Conversely, for a $(z, \alpha) \in epi\, h$, we have $\alpha \geq h(z) = f(w) + g(z - w)$ for a suitable $w \in X$. We may write $(z, \alpha) = (w, f(w)) + (z - w, \alpha - f(w))$, with $(w, f(w)) \in epi\, f$ and $(z - w, \alpha - f(w)) \in epi\, g$.   □

The function h that we defined in the above Lemma is called inf-convolution, and we will denote it by $f\#g$. Hence the Lemma asserts that the epigraph of the inf-convolution of two functions is the sum of their epigraphs. Let us observe that the inclusion $epi\, f + epi\, g \subset epi\,(f\#g)$ does not require for the inf to be attained.

The following two results are true in the general setting of Hilbert spaces, however for the sake of simplicity, we present easier proofs, derived from Lemma 2.8.

PROPOSITION 2.9. *Let* $S \subset \mathbf{R}^n$, *then* d_S *is convex provided that* S *is a closed convex set.*

PROOF. $d_S(x) = \inf_{w \in S} |x - w| = \inf_{y \in \mathbf{R}^n} (\delta_S(y) + |x - y|)$. We may invoke Lemma 2.8 since the inf is attained when S is closed, hence $epi\, d_S = epi\, \delta_S + epi\,(|.|)$. We conclude that $epi\, d_S$, and consequently d_S, are convex.   □

Let us observe that the convexity of S is also necessary since S is the set of minima of d_S and it is an exercise to prove that the set of minima of a convex function is convex. Another property of the distance function when S is a closed convex set is that the inf in its definition is attained at a single point.

PROPOSITION 2.10. *Let* $S \subset \mathbf{R}^n$ *be a closed convex set. Then for every* x, *there is a unique point* $\hat{x} \in S$ *such that* $d_S(x) = |x - \hat{x}|$ *(see Fig. 2.3).*

PROOF. As we said before, there is a point $\hat{x}$ such that $d_S(x) = |x - \hat{x}|$ since S is closed. If another point $\hat{y} \in S$ satisfies the same property, then all the points in the segment $[\hat{x}, \hat{y}]$, $t\hat{x} + (1 - t)\hat{y}$ satisfy the property too, since

$$d_S(x) \leq |t\hat{x} + (1 - t)\hat{y} - x|$$
$$\leq t|\hat{x} - x| + (1 - t)|\hat{y} - x|$$
$$= t d_S(x) + (1 - t) d_S(x) = d_S(x).$$

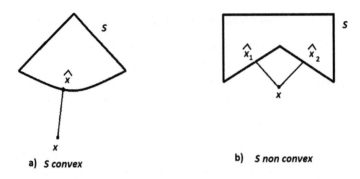

Fig. 2.3 Distance to a set S.

Therefore $[\hat{x}, \hat{y}]$ is contained in the sphere centered at x, with radius $d_S(x)$, which is impossible since no sphere in a Hilbert space can contain segments. □

Let us observe that we have required finite dimension in order to guarantee the existence of the closest point.

With respect to the definition of convex functions, it is interesting to observe that as a matter of fact, convexity of functions is a one-dimensional property, as a function is convex if and only if it is convex when restricted to every affine line in X. Taking advantage of this remark, it is easy to see that $|.|^2$ is a convex function since it is a non-negative polynomial of degree two when restricted to a one-dimensional line. We will come back to this property later.

An important property of convex functions is that their local minima are global. Let us prove it.

PROPOSITION 2.11. *Let $f : X \to (-\infty, +\infty]$ be a convex function. If $f(x_0) \le f(x)$ for every x in a V neighborhood of x_0, then the inequality holds for every $x \in X$.*

PROOF. Let us suppose that there exists a x_1 such that $f(x_1) < f(x_0)$. If $t \in (0, 1)$ is close enough to 1, then $tx_0 + (1 - t)x_1 \in V$ holds. On one hand $f(tx_0 + (1 - t)x_1) \ge f(x_0)$ since x_0 is a local minimum, but on the other $f(tx_0 + (1 - t)x_1) \le tf(x_0) + (1 - t)f(x_1) < tf(x_0) + (1 - t)f(x_0) = f(x_0)$. This contradiction proves that x_0 is a global minimum. □

2.2 CONTINUITY OF CONVEX FUNCTIONS

It is well known that discontinuous linear functionals on infinite dimensional spaces exist. In fact that property characterizes them. Convex functions however, have nice continuity properties. We will use the next Lemma to prove them.

LEMMA 2.12. *Let $x_0 \in X$, $r > 0$, and $f : B(x_0, r) \to \mathbf{R}$ be a convex function. Assume that $|f(x)| \le M$ for every $x \in B(x_0, r)$. Then $|f(x) - f(y)| \le \frac{4M}{r}|x - y|$ for every $x, y \in B\left(x_0, \frac{r}{2}\right)$.*

PROOF. Let x, y be two different points of $B\left(x_0, \frac{r}{2}\right)$, let us define $z = y + \frac{r}{2|y-x|}(y - x)$, clearly $z \in B(x_0, r)$. Moreover, $2|y - x|z = 2|y - x|y + r(y - x) = (2|y - x| + r)y - rx$, and therefore

$$y = \frac{2|y - x|}{2|y - x| + r}z + \frac{r}{2|y - x| + r}x$$

Consequently, y is a convex combination of x and z, and convexity of f gives

$$f(y) \le \frac{2|y - x|}{2|y - x| + r}f(z) + \frac{r}{2|y - x| + r}f(x)$$
$$= \frac{2|y - x|}{2|y - x| + r}(f(z) - f(x)) + f(x).$$

We deduce:

$$f(y) - f(x) \le \frac{2|y - x|}{2|y - x| + r}(f(z) - f(x)) \le \frac{2|y - x|}{2|y - x| + r}|f(z) - f(x)|$$
$$\le \frac{2|y - x|}{2|y - x| + r}2M \le \frac{2|y - x|}{r}2M = \frac{4M}{r}|y - x|.$$

Changing the roles of x and y yields the result.   □

The following result is an immediate consequence of this Lemma.

THEOREM 2.13. *Every convex, locally bounded function is locally Lipschitz.*

The fact that continuous functions are locally bounded, gives us the following corollary.

COROLLARY 2.14. *Every continuous convex function is locally Lipschitz.*

A discontinuous linear functional provides an example of a discontinuous convex function. However, when $X = \mathbf{R}^n$ those kind of functionals do not exist. In fact in this case, every convex function is continuous, and consequently locally Lipschitz. Before dealing with this nontrivial problem, we will give an elementary proof of the result when $n = 1$.

PROPOSITION 2.15. *A convex function $f : \mathbf{R} \to \mathbf{R}$ is continuous.*

PROOF. It is clear that it is enough to prove that f is continuous at 0 since convexity and continuity are stable under addition of constants to the argument. We first prove that $\lim_{x \to 0^+} f(x) = f(0)$.

We may write $x > 0$ as $x = (1 - x)0 + x1$, hence $f(x) \leq (1 - x)$ $f(0) + xf(1)$. Letting $x \to 0^+$ we get $\limsup_{x \to 0^+} f(x) \leq f(0)$. On the other hand $0 = \frac{x}{1+x}(-1) + \frac{1}{1+x}x$ and consequently $f(0) \leq \frac{x}{1+x}f(-1) + \frac{1}{1+x}f(x)$. Letting $x \to 0^+$ again, we get $f(0) \leq \liminf_{x \to 0^+} f(x)$, which finishes this case. The other case: $\lim_{x \to 0^-} f(x) = f(0)$ is left to the reader. $\qquad\square$

In order to prove the general case, we require a previous result which is important by itself, it is the so-called *Line Segment Principle*.

PROPOSITION 2.16. *Let $C \subset X$ be a convex set. Suppose that $x_0 \in \operatorname{int}C$, and $x_1 \in \overline{C}$, then the segment $(x_0, x_1) = \{(1 - t)x_0 + tx_1 : t \in (0, 1)\}$ lies on $\operatorname{int}C$.*

PROOF. Let $\varepsilon_0 > 0$ such that $x_0 + \varepsilon_0 B = B(x_0, \varepsilon_0) \subset C$. For every $\varepsilon_1 > 0$ we have that $(x_1 + \varepsilon_1 B) \cap C \neq \emptyset$, or $x_1 \in C + \varepsilon_1 B$ equivalently. Let $x_t = (1 - t)x_0 + tx_1 \in (x_0, x_1)$ be arbitrary, and define $\varepsilon_t = (1 - t)\varepsilon_0 - t\varepsilon_1$ with ε_1 small enough in order to guarantee that $\varepsilon_t > 0$. We claim that $x_t + \varepsilon_t B \subset C$ and consequently $x_t \in \operatorname{int}C$. Let us prove the claim:

$$
\begin{aligned}
x_t + \varepsilon_t B &= (1 - t)x_0 + tx_1 + \varepsilon_t B \subset (1 - t)x_0 + t(C + \varepsilon_1 B) + \varepsilon_t B \\
&= (1 - t)x_0 + (t\varepsilon_1 + \varepsilon_t)B + tC = (1 - t)x_0 + (1 - t)\varepsilon_0 B + tC \\
&= (1 - t)(x_0 + \varepsilon_0 B) + tC \subset (1 - t)C + tC = C,
\end{aligned}
$$

the last equality holds since C is convex. $\qquad\square$

An easy consequence of this proposition is that the interior of a convex set is convex, but other interesting consequences may be deduced.

PROPOSITION 2.17.   *If C is a convex set and $int\,C \neq \emptyset$, then*

$$int\,C = int\,\overline{C} \quad and \quad \overline{C} = \overline{int\,C}$$

PROOF.   The first equality just requires the proof of $int\,\overline{C} \subset int\,C$. Fix $x_0 \in int\,C$, for a $x \in int\,\overline{C}$, there exists a $\varepsilon > 0$ such that $B(x, \varepsilon) \subset \overline{C}$. For every $z \in B(x, \varepsilon)$, the segment $[x_0, z) \subset int\,C$, and in particular $[x_0, z_0) \subset int\,C$ where $z_0 = x + \frac{\varepsilon}{2|x-x_0|}(x - x_0)$ since $|z_0 - x| = \frac{\varepsilon}{2}$. This proves that $x \in [x_0, z_0) \subset int\,C$.

The other equality follows from the fact that for every $x \in \overline{C}$, the segment $[x_0, x) \subset int\,C$ by the Line Segment Principle (Proposition 2.16), and consequently $x \in \overline{int\,C}$.   □

Another consequence of the Line Segment Principle is that it allows us to characterize the interior of the epigraph of a convex function, but we require also the following geometrical Lemma. Its proof is left to the reader.

LEMMA 2.18.   *For every $x_0 \in \mathbf{R}^n$, there are $n + 1$ points $a_0, \ldots, a_n$ such that*

$$x_0 \in int\,(co\{a_0, \ldots, a_n\})$$

PROPOSITION 2.19.   *For a convex function $f : \mathbf{R}^n \to \mathbf{R}$ we have:*

$$int\,(epi f) = \{(x, r) : f(x) < r\}$$

PROOF.   If $(x_0, r_0) \in int\,(epi f)$ then $f(x_0) < r_0$ clearly since there is a $\varepsilon > 0$ such that $|x - x_0| < \varepsilon$ and $|r - r_0| < \varepsilon$, which implies $(x, r) \in epi f$, in particular $\left(x_0, r_0 - \frac{\varepsilon}{2}\right) \in epi f$, which implies $f(x_0) \leq r_0 - \frac{\varepsilon}{2} < r_0$.

Conversely, let (x_0, r_0) satisfy $f(x_0) < r_0$. As $x_0 \in \mathbf{R}^n$, we may suppose that it lies within the interior of a simplex $S = co\{a_0, \ldots, a_n\}$ generated by $n + 1$ points. If we define $r_1 = \max\{f(a_0), \ldots f(a_n), r_0\}$, we have that $f(x) \leq r_1$ for every $x \in S$, since x is a convex combination of $a_0, \ldots a_n$ and f is convex. So, the open set $int\,S \times (r_1, +\infty)$ is contained in $epi f$ and consequently in $int\,(epi f)$. If $r_2 > r_1$, we have $(x_0, r_2) \in int\,(epi f)$, and on the other hand, $(x_0, f(x_0)) \in epi f$, hence the open segment joining both points is contained in $int(epi f)$; as (x_0, r_0) lies between $(x_0, f(x_0))$ and (x_0, r_2) since $f(x_0) < r_0 \leq r_1 < r_2$, we conclude that $(x_0, r_0) \in int(epi f)$ too.   □

This result is no longer true for infinite dimensional Hilbert spaces. Let us consider an unbounded linear functional $f : X \to \mathbf{R}$, and let us observe that $epi f$ cannot contain any basic open set of $X \times \mathbf{R}$, that is

$$U \times I \not\subset epi f$$

for $U \subset X$ open ball, and $I \subset \mathbf{R}$ open interval, since f is unbounded on U.

COROLLARY 2.20.   *Every convex function $f : \mathbf{R}^n \to \mathbf{R}$ is usc.*

PROOF.   It is enough to observe that $hypo f = (X \times \mathbf{R}) \backslash \{(x, r) : f(x) < r\}$ is closed.   □

COROLLARY 2.21.   *Every convex function $f : \mathbf{R}^n \to \mathbf{R}$ is lsc.*

PROOF.   Let $(x, r) \in \overline{epi f}$. If $s > \max\{f(x), r\}$, we have that $(x, s) \in int(epi f)$, by Proposition 2.19. The open segment $((x, s), (x, r))$ lies within $int(epi f)$ by Proposition 2.16, but this means that $f(x) < t$ for every $t > r$ by Proposition 2.19 again, hence $f(x) \leq r$ and consequently $(x, r) \in epi f$. We conclude that $epi f$ is closed and therefore f is *lsc*.   □

Both of these corollaries together give us:

THEOREM 2.22.   *Every convex function $f : \mathbf{R}^n \to \mathbf{R}$ is continuous.*

We have seen that Moreau envelopes regularize discontinuities of *lsc* functions. If we are dealing with convex functions the regularization is better. The following theorem, which holds in the more general setting of Hilbert spaces, asserts that the Moreau envelope of a convex function is C^1.

THEOREM 2.23.   *Let $f : \mathbf{R}^n \to (-\infty, +\infty]$ be a lsc convex function bounded below, then the following properties hold:*

(i)   *For every $x \in X$, there is a unique x_0 such that $f_\lambda(x) = f(x_0) + \frac{1}{2\lambda}|x - x_0|^2$.*

(ii)   *f_λ is convex.*

(iii)   *f_λ is continuously differentiable, and $\nabla f_\lambda(x) = \frac{1}{\lambda}(x - x_0)$.*

PROOF. Let $b = \inf f$. First of all we are going to prove that the inf in the definition of f_λ is attained.

From the inf definition we know that there are vectors w_n such that $f(w_n) + \frac{1}{2\lambda}|x - w_n|^2 < f_\lambda(x) + \frac{1}{n}$, from which we deduce

$$b + \frac{1}{2\lambda}|x - w_n|^2 < f_\lambda(x) + \frac{1}{n} \le f(x) + \frac{1}{n}$$

and consequently

$$|x - w_n|^2 \le 2\lambda(f(x) + 1 - b)$$

hence, by compactness of closed balls in $\mathbf{R}^n$, we may assume that the bounded sequence $\{w_n\}$ converges to a point x_0 that satisfies $f(x_0) + \frac{1}{2\lambda}|x - x_0|^2 = f_\lambda(x)$ because f is *lsc*. We have proved not only that the inf is attained, but that every minimizing sequence is bounded.

Once we have established the hypotheses of Lemma 2.8, it holds that $epi f_\lambda = epi f + epi\left(\frac{1}{2\lambda}|.|^2\right)$ is convex since it is the sum of two convex sets. This proves that f_λ is convex.

Let us suppose that the inf is attained at two different points x_1, x_2. Then as a consequence of the Parallelogram Law, we deduce the following strict inequality

$$f_\lambda(x) \le f\left(\frac{1}{2}(x_1 + x_2)\right) + \frac{1}{2\lambda}|x - \frac{1}{2}(x_1 + x_2)|^2$$

$$\le \frac{1}{2}(f(x_1) + f(x_2)) + \frac{1}{2\lambda}|\frac{1}{2}(x - x_1) + \frac{1}{2}(x - x_2)|^2 < \frac{1}{2}(f(x_1)$$

$$+ f(x_2)) + \frac{1}{2\lambda}\left(\frac{1}{2}|x - x_1|^2 + \frac{1}{2}|x - x_2|^2\right)$$

$$= \frac{1}{2}f_\lambda(x) + \frac{1}{2}f_\lambda(x) = f_\lambda(x)$$

This contradiction proves (i).

Before proving part (iii), we are going to see that the unique point x_0 such that $f_\lambda(x) = f(x_0) + \frac{1}{2\lambda}|x - x_0|^2$, depends on x continuously. Let us consider a sequence $\{x^n\}$ converging to x. Each x^n has a corresponding x_0^n, and x_0 corresponds to x. Does $\lim x_0^n = x_0$?

$\{f_\lambda(x^n)\}$ converges to $f_\lambda(x)$ since f_λ, which is finite and convex, is continuous, hence that sequence is bounded. An argument similar to the one at the beginning of the proof guarantees that the sequence $\{x_0^n\}$ is bounded, so we may assume that it converges to a point, namely $\hat{x}_0$. We have

$$f(\hat{x}_0) + \frac{1}{2\lambda}|x - \hat{x}_0|^2 \le \liminf \left(f(x_0^n) + \frac{1}{2\lambda}|x^n - x_0^n|^2 \right)$$
$$= \liminf f_\lambda(x^n) = f_\lambda(x)$$

where the inequality follows from the fact that f is *lsc*, while the last equality holds since f_λ is continuous. Uniqueness of x_0 proves that $x_0 = \hat{x}_0$, and the work is done.

Now, we establish the gradient formula. We have to check that

$$\lim_{h \to 0} \frac{f_\lambda(x + h) - f_\lambda(x) - \langle \frac{1}{\lambda}(x - x_0), h \rangle}{|h|} = 0$$

but this is equivalent to proving that $\nabla g(0) = 0$ where $g(h) = f_\lambda(x + h) - f_\lambda(x) - \langle \frac{1}{\lambda}(x - x_0), h \rangle$.

We have

$$g(h) = f_\lambda(x + h) - \left(f(x_0) + \frac{1}{2\lambda}|x - x_0|^2 \right) - \left\langle \frac{1}{\lambda}(x - x_0), h \right\rangle$$
$$\le f(x_0) + \frac{1}{2\lambda}|x + h - x_0|^2 - \left(f(x_0) + \frac{1}{2\lambda}|x - x_0|^2 \right)$$
$$- \left\langle \frac{1}{\lambda}(x - x_0), h \right\rangle = \frac{1}{2\lambda}|h|^2$$

On the other hand, $0 = g(0) = g(\frac{1}{2}h + \frac{1}{2}(-h)) \le \frac{1}{2}(g(h) + g(-h))$ since g is convex, hence $g(h) \ge -g(-h) \ge -\frac{1}{2\lambda}|h|^2$. Both inequalities together give us $|g(h)| \le \frac{1}{2\lambda}|h|^2$ and consequently $\nabla g(0) = 0$. Once the formula has been established, the continuity of x_0 with respect to x in the expression $f_\lambda(x) = f(x_0) + \frac{1}{2\lambda}|x - x_0|^2$ guarantees that f_λ is continuously differentiable. $\square$

2.3 SEPARATION RESULTS

We will now study the problem of separating convex sets by means of hyper-planes. The following results are true in a more general setting, in particular for Hilbert spaces, but for the sake of simplicity we will present them when $X = \mathbf{R}^n$.

THEOREM 2.24. *(Minkowski Separation Theorem).* *Let* $C_1, C_2 \subset \mathbf{R}^n$ *be nonempty and convex. If* $0 \notin int(C_1 - C_2)$, *then there exist* $\alpha \in \mathbf{R}$, $v \in \mathbf{R}^n$ *such that*

$$C_1 \subset \{x : \langle x, v \rangle \le \alpha\}, \quad C_2 \subset \{x : \langle x, v \rangle \ge \alpha\}.$$

PROOF. Let $C = \overline{C_1 - C_2}$. C is convex since it is the closure of the convex set $C_1 - C_2$. We claim that there exists a nonzero vector v such that $\langle x, v \rangle \le 0$ for every $x \in C$. If this is the case, then $\langle x_1, v \rangle \le \langle x_2, v \rangle$ for every $x_1 \in C_1$, $x_2 \in C_2$. A real number $\alpha \in [\sup_{x_1 \in C_1} \langle x_1, v \rangle, \inf_{x_2 \in C_2} \langle x_2, v \rangle]$, together with v, satisfies the required condition.

Let us prove the claim. If $0 \notin C$, let $x_0 \in C$ be the unique point of C nearest to 0. Let $x \in C$, the segment $[x, x_0]$ lies in C too, therefore the function $g(t) = |(1 - t)x_0 + tx|^2$ satisfies $g(t) > g(0)$ for every $t \in (0, 1)$, and consequently $0 \le g'(0) = 2\langle x_0, x - x_0 \rangle$. Hence we may take $v = -x_0$, since $\langle x, x_0 \rangle \ge 0$ for every $x \in C$.

Let us suppose now that $0 \in C$, by Proposition 2.17 we know that $0 \notin intC$ because $0 \notin int(C_1 - C_2)$. The fact that $0 \notin intC$ implies that there is a sequence $\{x_n\}$ converging to 0, such that $x_n \notin C$. Now, we have that 0 does not belong to the convex $-x_n + C$, and the same argument as before gives us a nonzero vector v_n such that

$$\langle v_n, -x_n + x \rangle \le 0 \quad \text{for every } x \in C. \tag{2.3}$$

We may assume without loss of generality that $|v_n| = 1$, and also that $\{v_n\}$ converges to a vector v by compactness of the unit ball in $\mathbf{R}^n$. Taking limits in (2.3), we have $\langle v, x \rangle \le 0$. $\qquad\square$

Let us observe that the condition $0 \notin int(C_1 - C_2)$ is weaker than $0 \notin C_1 - C_2$ which is equivalent to $C_1 \cap C_2 = \emptyset$. Nevertheless we can say more.

COROLLARY 2.25. *Let* $C_1, C_2 \subset \mathbf{R}^n$ *be two nonempty convex sets, assume that* $intC_1 \ne \emptyset$ *and that* $intC_1 \cap C_2 = \emptyset$. *Then there are a nonzero vector* v *and a real number* α *such that*

$$\langle v, x_1 \rangle \le \alpha \quad for \; every \;\; x_1 \in C_1 \quad and \quad \langle v, x_2 \rangle \ge \alpha \quad for \; every \; x_2 \in C_2.$$

PROOF. It is enough to observe that $0 \notin int(C_1 - C_2)$ which is open. This condition guarantees the existence of $\alpha \in \mathbf{R}$ and $v \in \mathbf{R}^n$, $v \ne 0$ such that $\langle v, x_1 \rangle \leq \alpha$ for every $x_1 \in int C_1$, and $\langle v, x_2 \rangle \geq \alpha$ for every $x_2 \in C_2$. But $\langle v, x_1 \rangle \leq \alpha$ for every $x_1 \in \overline{int C_1} = \overline{C_1}$ also by continuity.   □

One remark before proving the next Corollary: what are convex sets with an empty interior like? It is not difficult to see that a convex set $C \subset \mathbf{R}^n$ such that $int C = \emptyset$ is necessarily contained in an affine subspace L with $dim L < n$, since otherwise it would contain a simplex with nonempty interior.

COROLLARY 2.26.   *Let C be a nonempty convex subset of $\mathbf{R}^n$, let $a \in \partial C$. There exists a nonzero $v \in \mathbf{R}^n$ such that $\langle v, x \rangle \leq \langle v, a \rangle$ for every $x \in C$.*

PROOF.   If $int C = \emptyset$, the remark above implies that C, and consequently $\overline{C}$ are contained in an affine subspace L, hence there is a nonzero vector v such that $\langle v, x \rangle = \alpha$ for every $x \in \overline{C}$. If $int C \neq \emptyset$, we invoke Corollary 2.25, with $C_1 = C$ and $C_2 = \{a\}$, to deduce that there exist $v \in \mathbf{R}^n$ and $\alpha \in \mathbf{R}$ such that

$$\langle v, x \rangle \leq \alpha \quad \text{for every} \quad x \in C \qquad \langle v, a \rangle \geq \alpha.$$

The fact that $a \in \partial C$ implies $\langle v, a \rangle \leq \alpha$ also, hence $\langle v, a \rangle = \alpha$.   □

A vector v that satisfies the above property defines a functional $\phi_v(x) = \langle x, v \rangle$ called a support functional of C at a, and the hyperplane $\langle x, v \rangle = \alpha$ is called a support hyperplane. Sometimes we will refer to v as a support of C at x.

COROLLARY 2.27.   *Every closed convex subset of $\mathbf{R}^n$ is the intersection of closed half spaces.*

PROOF.   For $x_0 \notin C$ let $\varepsilon > 0$ be such that $B(x_0, \varepsilon) \cap C = \emptyset$. We invoke Minkowski's Separation Theorem with $C_1 = C$ and $C_2 = B(x_0, \varepsilon)$, and deduce the existence of $v \in \mathbf{R}^n$ and $\alpha \in \mathbf{R}$ such that:

$$\langle v, x_1 \rangle \leq \alpha \quad \text{for every} \quad x \in C \quad \text{and} \quad \langle v, y \rangle \geq \alpha \quad \text{for every} \quad y \in B(x_0, \varepsilon).$$

The second inequality implies $\langle v, x_0 \rangle > \alpha$. Hence x_0 does not belong to the half space $\langle x, v \rangle \leq \alpha$. We have proved that for every $x_0 \notin C$ there is a half space that contains C but does not contain x_0. This implies that C is the intersection of all half spaces containing it.   □

Taking into account that the converse is trivially true, this property characterizes convexity. As we said before all these results are true in a more general setting, namely in Banach spaces; the separation result in that case is known as the Hahn-Banach Theorem.

2.4 CONVEXITY AND DIFFERENTIABILITY

Finally, we look at convexity under a different scope. The aim of the following lines is to study convexity when the involved function is differentiable. We will start with a well-known one-dimensional result, and after that we will introduce another, probably not so well-known, higher dimensional result.

PROPOSITION 2.28. *Let $I \subset \mathbf{R}$ be an open interval and $f : I \to \mathbf{R}$ a differentiable function. The following asserts are equivalent:*

(i) f is convex.
(ii) f' is nondecreasing.
(iii) $f(y) \geq f(x) + f'(x)(y - x)$ for every $x, y \in I$.
(iv) $f''(x) \geq 0$ for every $x \in I$ (for this assert we also require f to be twice differentiable).

PROOF. If f is convex, it is easy to see that for $x, y, z \in I, x < y < z$,

$$\frac{f(y) - f(x)}{y - x} \leq \frac{f(z) - f(x)}{z - x} \leq \frac{f(z) - f(y)}{z - y} \tag{2.4}$$

It is enough to observe that $y = \frac{y-x}{z-x}z + \frac{z-y}{z-x}x$ and consequently

$$f(y) \leq \frac{y - x}{z - x} f(z) + \frac{z - y}{z - x} f(x)$$

from where

$$f(y) - f(x) \leq \frac{y - x}{z - x} f(z) + \frac{z - y - z + x}{z - x} f(x) = \frac{y - x}{z - x}(f(z) - f(x)),$$

which gives us the first inequality. The second one is proved in a similar way. Once we have established (2.4), letting $y \to x$ and $y \to z$ successively, we get

$$f'(x) \leq \frac{f(z) - f(x)}{z - x} \leq f'(z).$$

This proves that f' is nondecreasing.

In order to prove (iii), let us consider the function $g_x(y) = f(y) - f(x) - f'(x)(y-x)$. We have that $g'_x(y) = f'(y) - f'(x)$, hence $g'_x(y) \geq 0$ if $y \geq x$ and $g'_x(y) \leq 0$ if $y \leq x$. We deduce that g'_x attains a minimum at $y = x$, that is $0 = g_x(x) \leq g_x(y) = f(y) - f(x) - f'(x)(y-x)$.

Now, we assume (iii). Given $x, y \in I$ and $t \in (0, 1)$, we have:

$$f(y) \geq f(tx + (1-t)y) + f'(tx + (1-t)y)(y - tx - (1-t)y)$$
$$= f(tx + (1-t)y) + f'(tx + (1-t)y)t(y - x)$$
$$f(x) \geq f(tx + (1-t)y) + f'(tx + (1-t)y)(x - tx - (1-t)y))$$
$$= f(tx + (1-t)y) + f'(tx + (1-t)y)(1-t)(x - y)$$

multiplying the first inequality by $(1-t)$, the second by t, and adding them, we obtain $f(tx + (1-t)y) \leq tf(x) + (1-t)f(y)$.

Finally, if f is twice differentiable we know, from elementary calculus, that f' is nondecreasing if and only if $f''(x) \geq 0$ for every $x \in I$.   □

We proceed to extend the above result to higher dimensions.

THEOREM 2.29. *Let $G \subset X$ be an open convex set and $f : G \to \mathbf{R}$ a differentiable function. Then the following asserts are equivalent:*

(i) f is convex.
(ii) $\langle y - x, \nabla f(y) - \nabla f(x) \rangle \geq 0$ for every $x, y \in G$.
(iii) $f(x) \geq f(y) + \langle \nabla f(y), x - y \rangle$ for every $x, y \in G$.

PROOF. Let us suppose that f is convex. For $x, y \in G$, we define the convex function $g : I \to \mathbf{R}$ by $g(t) = f(tx + (1-t)y)$, where I is an open interval, that contains $[0, 1]$, and such that $tx + (1-t)y \in G$ for every $t \in I$. The existence of such an interval is an easy consequence of the fact that G is convex and open. From the Chain Rule, we have that $g'(t) = \langle \nabla f(tx + (1-t)y), x - y \rangle$. Assert (ii) in Proposition 2.28 implies $g'(1) \geq g'(0)$ from where we deduce:

$$\langle \nabla f(x), x - y \rangle \geq \langle \nabla f(y), x - y \rangle$$

In order to deduce (iii) from (ii) it is enough to define the function

$$g(t) = f(tx + (1-t)y).$$

Again, if $t \geq s$ we have that

$$g'(t) - g'(s) = \langle \nabla f(tx + (1-t)y), x - y \rangle - \langle \nabla f(sx + (1-s)y), x - y \rangle$$
$$= \langle \nabla f(tx + (1-t)y) - \nabla f(sx + (1-s)y), x - y \rangle.$$

From this formula we deduce:

$$(t-s)(g'(t) - g'(s)) = \langle \nabla f(tx + (1-t)y)$$
$$- \nabla f(sx + (1-s)y), (t-s)(x-y) \rangle$$
$$= \langle \nabla f(tx + (1-t)y) - \nabla f(sx + (1-s)y),$$
$$tx + (1-t)y - sx - (1-s)y \rangle \geq 0,$$

and consequently $g'(t) \geq g'(s)$, that is g' is nondecreasing. Then we have $g(1) \geq g(0) + g'(0)$, by Proposition 2.28, which implies $f(x) \geq f(y) + \langle \nabla f(y), x - y \rangle$.

It remains to prove that (iii) implies (i). As already noted, the function f is convex if and only if it is convex when restricted to lines contained in G. For this, it is enough to prove that the function $g(t) = f(y + tz)$, where $y \in G$ and $z \in X$, is convex in its domain, which is an open interval that contains 0. For $t, s \in domg$,

$$g(s) + g'(s)(t-s) = f(y+sz) + \langle \nabla f(y+sz), z \rangle (t-s)$$
$$= f(y+sz) + \langle \nabla f(y+sz), z \rangle (t-s)$$
$$= f(y+sz) + \langle \nabla f(y+sz), y + tz - (y+sz) \rangle$$
$$\leq f(y+tz) = g(t)$$

hence g is convex by Proposition 2.28.   □

COROLLARY 2.30. *Let $G \subset \mathbf{R}^n$ be an open convex set. A twice differentiable function $f : G \to \mathbf{R}$ is convex if and only if the Hessian matrix $\nabla^2 f(x)$ is positive semi-definite for every $x \in G$.*

PROOF. As in Proposition 2.28 it is enough to characterize when the function $g(t) = f(y + tz)$, for $y \in G$ and $z \in \mathbf{R}^n$, is convex. As $g'(t) = \langle \nabla f(y+tz), z \rangle$, we have that $g''(t) = z \nabla^2 f(y+tz)\overline{z}$. If $\nabla^2 f(x)$ is positive semi-definite for every $x \in G$, we have $g''(t) \geq 0$ for every function g and every $t \in domg$ and therefore f is convex. Conversely, if f is convex, we have $g''(t) \geq 0$, $g''(0) \geq 0$ in particular, which implies that $\nabla^2 f(y)$ are positive semi-definite for every $y \in G$.   □

2.5 PROBLEMS

(1) Prove Proposition 2.1.

(2) Give an example of a convex set C, and two real numbers λ_1, λ_2 such that $\lambda_1 C + \lambda_2 C \neq (\lambda_1 + \lambda_2)C$.

(3) Give an example of a nonconvex set A such that $A + A \neq 2A$.

(4) Prove that for every $x_0 \in \mathbf{R}^n$ there are $n + 1$ points $a_0, \ldots, a_n$ such that
$$x_0 \in int\left(co\{a_0, \ldots, a_n\}\right)$$

(5) Prove that $co(A + B) = coA + coB$.

(6) Find a closed set whose convex hull is not closed.

(7) Prove that $\overline{co}(A) + \overline{co}(B) \subset \overline{co}(A + B)$.

(8) Find a couple of subsets A and B satisfying $\overline{co}(A) + \overline{co}(B) \neq \overline{co}(A + B)$.

(9) Prove that $\overline{co}(A) + \overline{co}(B) = \overline{co}(A + B)$ provided that A is compact.

(10) Describe the convex hull of a subset A of $\mathbf{R}$. (No more than 13 characters should be required).

(11) Prove that for every $A \subset \mathbf{R}^2$ it is enough to take convex combinations of three points in order to define coA, but there are sets for which two points are not sufficient. (Hint: Translate the result to an elementary assert on polygons.)

(12) Prove that if all the functions $f_i : X \to (\infty, +\infty]$ are convex and $\lambda_i \geq 0$ for $i = 1 \ldots m$, then $\sum_{i=1}^{m} \lambda_i f_i$ is convex too.

(13) Give an example of two convex functions $f, g : X \to (-\infty, +\infty]$, such that $f.g$ is not convex.

(14) Is f^2 convex for every convex function f?

(15) Draw the epigraph of the distance function d_S, where $S \subset \mathbf{R}$ is the following set: $S = \{\frac{1}{n} : n \in \mathbb{N}\} \cup \{0\}$.

(16) Draw the graph of the distance function d_S, where
$$S = \{(x, y) \in \mathbf{R}^2 : \min\{x, y\} \leq 0\}.$$

(17) Prove that the set of minima of a convex function is necessarily convex.

(18) Prove that the inf-convolution is commutative, that is $f\#g = g\#f$.

(19) What is the inf-convolution of two functions if one of them is constant?

(20) Let us consider $f, g : \mathbf{R} \to \mathbf{R}$ defined by $f(t) = e^t$ and $g(t) = 0$. Prove that $epi(f\#g) \neq epif + epig$.

(21) Prove that a convex bounded function $f : \mathbf{R}^n \to \mathbf{R}$ is necessarily constant.

(22) Give an example of a convex set C, and two points $x_0 \in C$ and $x_1 \in \overline{C}$ such that $(x_0, x_1) \not\subset C$ (i.e. the assumption $x_0 \in int\,C$ cannot be relaxed in the Line Segment Principle, 2.16).

(23) Check all the asserts in Theorem 2.23 for the Moreau envelope, $f_{\frac{1}{2}}$, of the function $f : \mathbf{R} \to \mathbf{R}$ defined by $f(t) = |t|$.

(24) Let $f, g : \mathbf{R}^n \to \mathbf{R}$ be such that $f \le g$. Let us assume that f is concave and g convex. Prove that there is an affine function L such that $f \le L \le g$.

(25) Let $g : \mathbf{R}^n \to (-\infty, +\infty]$, we define

$$f(x) = \inf\{\lambda : (x, \lambda) \in co(epi\,g)\}$$

Assume that $f(x) > -\infty$ for every x, which happens if g is bounded below for instance. Prove that f is the greatest convex function satisfying $f \le g$. This function is called *convexification* or *convex hull* of g.

(26) Calculate the convexification of the function $f : \mathbf{R} \to \mathbf{R}$ defined by

$$f(t) = \begin{cases} 0 & \text{if } t \le 0, \\ \frac{t}{t^2+1} & t > 0. \end{cases}$$

(27) Prove that the epigraph of the convexification of a function f is the closed convex hull of $epi\,f$, $\overline{co}(epi\,f)$.

CHAPTER 3

The Subdifferential of a Convex Function

Throughout this chapter $f : X \to (-\infty, +\infty]$ will be a *lsc* convex function. When $dom f = X = \mathbf{R}^n$, convexity guarantees continuity. From the precedent results we know that $epi f$ is a closed convex set. We start with a definition.

DEFINITION 3.1. Let $f : X \to (-\infty, +\infty]$, for a point $x \in dom f$, we define the subdifferential of f at x, $\partial f(x)$ as

$$\partial f(x) = \{\zeta \in X : f(y) \geq f(x) + \langle \zeta, y - x \rangle \text{ for every } y \in X\}$$

Sometimes the following alternative notation is used: $D^- f(x)$ or $\partial^- f(x)$ instead of $\partial f(x)$. This is a way of remarking the "sub" character of the definition in contraposition to the superdifferential that we will introduce later. Let us observe that $\partial f(x)$ is defined as a set. The elements of $\partial f(x)$ are usually called subgradients of f at x. The following example proves that, generally, the subdifferential of a function is a set valued function.

Example. Given $f : \mathbf{R} \to \mathbf{R}$ defined by $f(x) = |x|$, it is left as an exercise to the reader to see that $\partial f(x) = \pm 1$ depending on whether $x > 0$ or $x < 0$, while $\partial f(0) = [-1, 1]$ (see Fig. 3.1).

Our next proposition is very important despite its simplicity, because it motivates the importance of the concept that we have just introduced.

PROPOSITION 3.2. *If f attains a minimum at x, then $0 \in \partial f(x)$.*

PROOF. The minimum is global since f is convex, hence $f(y) \geq f(x) = f(x) + \langle 0, y - x \rangle$ for every $y \in X$. □

3.1 SUBDIFFERENTIAL PROPERTIES

We list some elementary properties of the subdifferential.

PROPOSITION 3.3. *$\partial f(x)$ is a closed convex set for every $x \in X$.*

PROOF. If $f(y) \geq f(x) + \langle \zeta_1, y - x \rangle$ and $f(y) \geq f(x) + \langle \zeta_2, y - x \rangle$ for every $y \in X$, it is immediate to see that $f(y) \geq f(x) + \langle t\zeta_1 + (1-t)\zeta_2, y - x \rangle$

An Introduction to Nonsmooth Analysis. http://dx.doi.org/10.1016/B978-0-12-800731-0.00003-5

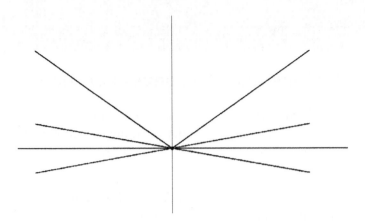

Fig. 3.1 Subdifferential of the absolute value function.

for every $t \in (0, 1)$. This proves convexity. Proving that it is a closed set is left to the reader. □

What is the meaning of a subgradient? If we consider the affine function $g(y) = f(x) + \langle \zeta, y - x \rangle$ for a $\zeta \in \partial f(x)$, it results that $f(y) \geq g(y)$ for every $y \in X$. Geometrically this means that $epi f$ is "above" the affine hyperplane consisting of the graph of g, but considering that $f(x) = g(x)$ we have that this hyperplane is "tangent" to the convex set $epi f$, and moreover, $epi f$ lies in one of the half spaces defined by the graph of g. We have in particular the following proposition.

PROPOSITION 3.4. *If $x \in int(dom f)$, and f is differentiable at x, then $\partial f(x) = \{\nabla f(x)\}$.*

PROOF. In order to verify $\nabla f(x) \in \partial f(x)$, we have to prove $f(y) \geq f(x) + \langle \nabla f(x), y - x \rangle$ for every $y \in X$. We suppose that there exists $h \in X$ such that $f(x + h) - f(x) - \langle \nabla f(x), h \rangle < 0$ and we will arrive at a contradiction.

Let us define $g(t) = f(x + th) - f(x) - t\langle \nabla f(x), h \rangle$, we have that $g(0) = 0$ and $g(1) < 0$. For every $t \in (0, 1)$ convexity of g implies that $g(t) \leq (1 - t)g(0) + tg(1) = tg(1)$, hence

$$0 = \lim_{t \downarrow 0} \frac{f(x + th) - f(x) - \langle \nabla f(x), th \rangle}{t} = \lim_{t \downarrow 0} \frac{g(t)}{t} \leq g(1) < 0$$

On the other hand, if $\zeta \in \partial f(x)$, we have

$$\left\langle \zeta - \nabla f(x), \frac{h}{|h|} \right\rangle = \frac{f(x+h) - f(x) - \langle \nabla f(x), h \rangle}{|h|}$$
$$- \frac{f(x+h) - f(x) - \langle \zeta, h \rangle}{|h|}$$
$$\leq \frac{f(x+h) - f(x) - \langle \nabla f(x), h \rangle}{|h|}$$

hence

$$\limsup_{h \to 0} \left\langle \zeta - \nabla f(x), \frac{h}{|h|} \right\rangle \leq \lim_{h \to 0} \frac{f(x+h) - f(x) - \langle \nabla f(x), h \rangle}{|h|} = 0$$

which implies $\langle \zeta - \nabla f(x), v \rangle \leq 0$ for every $v \in \mathbf{R}^n$ with $|v| = 1$ and therefore for every $v \in \mathbf{R}^n$. We conclude that $\zeta = \nabla f(x)$ necessarily. $\square$

The proposition that we present now is specific to convex functions. It is true for Hilbert spaces X in general, but in order to keep the book self-contained we will only prove it for $X = \mathbf{R}^n$. The clue of the simpler proof is Minkowski's Separation Theorem. For the more general case, we would have to invoke Hahn-Banach's Theorem.

PROPOSITION 3.5. *Let $f : \mathbf{R}^n \to (-\infty, +\infty]$ be a lsc convex function, then $\partial f(x)$ is nonempty for every $x \in int(dom f)$*

PROOF. Let $x \in dom f$. The set $epi f$ is a closed convex subset of $\mathbf{R}^{n+1}$, and $(x, f(x)) \in \partial(epi f)$. Corollary 2.26 tells us that there is a nonzero $(v, r) \in \mathbf{R}^{n+1}$ such that

$$\langle (v, r), (y, t) \rangle \leq \langle (v, r), (x, f(x)) \rangle \text{ for every } (y, t) \in epi f. \tag{3.1}$$

We have in particular:

$$\langle (v, r), (y, f(y)) \rangle \leq \langle (v, r), (x, f(x)) \rangle \text{ for every } y \in dom f.$$

From this inequality we deduce the following one: $\langle v, y - x \rangle \leq r(f(x) - f(y))$ for every $y \in dom f$. The real number r cannot be 0 since this would imply $\langle v, y - x \rangle \leq 0$ for every $y \in dom f$, for every y in an open ball around x in particular. This implies that $\langle v, z \rangle \leq 0$ for every z in an open ball around the origin, and consequently $v = 0$ which is not possible since (v, r) is a nonzero vector.

Once established that $r \neq 0$, let us observe that r cannot be positive either, since taking $t \to +\infty$ in (3.1) would lead to a contradiction. We conclude that $r < 0$, and consequently $\langle v, y - x \rangle \leq -r(f(y) - f(x))$, hence $f(x) + \langle -\frac{1}{r}v, y - x \rangle \leq f(y)$ and $-\frac{1}{r}v \in \partial f(x)$. $\qquad\square$

Let us observe that the proof of Proposition 3.5 involves two interesting geometrical facts. Firstly, $r < 0$ means that the vectors that define the support of the convex epigraph point downwards. Secondly, that although there are "many" vectors (v, r) that define a given support, namely $\lambda(v, r)$, the quotient $\frac{-\lambda v}{\lambda r}$ remains constant.

The next example, apart of its intrinsic interest, proves that $\partial f(x)$ can be nonempty for points belonging to $\partial(domf)$.

Example. Let $S \subset X$ be closed and convex. The indicator function δ_S is *lsc*, convex and $dom(\delta_S) = S$. If $x \in intS$, then $\partial(\delta_S)(x) = \{0\}$. If $x \in \partial S$ a bit more work is required. $\zeta \in \partial(\delta_S)(x)$ if and only if $\delta_S(y) \geq \langle \zeta, y - x \rangle$ if and only if $\langle \zeta, y - x \rangle \leq 0$ for every $y \in S$. The fact that S is convex guarantees the existence of those ζ, as they are the supports of S at x.

Let us observe that for real-valued convex functions $f : \mathbf{R}^n \to \mathbf{R}$ we have that $\partial f(x) \neq \emptyset$ for every $x \in \mathbf{R}^n$. This observation has the following interesting consequence.

PROPOSITION 3.6.   *Every convex function* $f : \mathbf{R}^n \to \mathbf{R}$ *is the supreme of a family of affine functions.*

PROOF.   For every $x \in \mathbf{R}^n$ and every $\zeta \in \partial f(x)$ we consider the affine function $g_{x,\zeta}(y) = f(x) + \langle \zeta, y - x \rangle$. We claim that $f = \sup\{g_{x,\zeta} : \zeta \in \partial f(x), x \in \mathbf{R}^n\}$, equivalently $epif = \cap_{x,\zeta}epi(g_{x,\zeta})$. In order to prove this equality it is enough to observe that $epif \subset epi(g_{x,\zeta})$ for every $\zeta \in \partial f(x)$ and $x \in \mathbf{R}^n$, while if $(x, t) \notin epif$, then $(x, t) \notin epi(g_{x,\zeta})$ for any $\zeta \in \partial f(x)$ since $g_{x,\zeta}(x) = f(x)$. $\qquad\square$

Our next goal is to find equivalent conditions for subdifferentiability of a function. We list these conditions in the following theorem.

THEOREM 3.7.   *For a lsc convex function* $f : X \to (-\infty, +\infty]$, *and a point* $x \in domf$, *the following asserts are equivalent:*

(i) $\zeta \in \partial f(x)$.

(ii)   There exists a C^1 function $\varphi : X \to \mathbf{R}$ such that $\varphi \le f$, $\varphi(x) = f(x)$, and $\nabla\varphi(x) = \zeta$.

(iii) There exists a C^1 function $\varphi : X \to \mathbf{R}$ such that $\varphi(x) = f(x)$, $\nabla\varphi(x) = \zeta$, and $\varphi \le f$ in a neighborhood of x.

(iv)   There exists a differentiable function $\varphi : X \to \mathbf{R}$ such that $\varphi(x) = f(x)$, $\nabla\varphi(x) = \zeta$, and $\varphi \le f$ in a neighborhood of x.

(v) $\liminf_{h \to 0} \frac{f(x+h)-f(x)-\langle \zeta, h\rangle}{|h|} \ge 0$. (Frechet subdifferential)

(vi)   There exists a C^2 function $\varphi : X \to \mathbf{R}$ such that $\varphi(x) = f(x)$, $\nabla\varphi(x) = \zeta$, and $\varphi \le f$ in a neighborhood of x.

(vii) There exist $r > 0$, $\sigma > 0$ such that $f(y) \ge f(x) + \langle \zeta, y - x\rangle - \sigma|y - x|^2$ for every $y \in B(x, r)$. (Proximal subdifferential)

PROOF. (i) $\Rightarrow$ (ii). The affine function $f(x) + \langle \zeta, y - x\rangle$ satisfies the required conditions.

(ii) $\Rightarrow$ (iii). Trivial.

(iii) $\Rightarrow$ (iv). Trivial.

(iv) $\Rightarrow$ (v). We have that

$$\frac{f(x + h) - f(x) - \langle \zeta, h\rangle}{|h|} \ge \frac{\varphi(x + h) - \varphi(x) - \langle \nabla\varphi(x), h\rangle}{|h|}$$

for h small enough. Hence

$$\liminf_{h \to 0} \frac{f(x + h) - f(x) - \langle \zeta, h\rangle}{|h|}$$
$$\ge \lim_{h \to 0} \frac{\varphi(x + h) - \varphi(x) - \langle \nabla\varphi(x), h\rangle}{|h|} = 0$$

as φ is differentiable.

(v) $\Rightarrow$ (i). Let us suppose that there exists an h such that $f(x + h) - f(x) - \langle \zeta, h\rangle < 0$. We consider the convex function $g(t) = f(x + th) - f(x) - t\langle \zeta, h\rangle$ for $t \in \mathbf{R}$. It holds that $g(0) = 0$ and $g(1) < 0$, and necessarily $g(t) \le (1 - t)g(0) + tg(1) = tg(1)$ for every $t \in [0, 1]$ by convexity. Hence $\liminf_{t \downarrow 0} \frac{g(t)}{t} \le g(1)$ and consequently $\liminf_{h \to 0} \frac{f(x+h)-f(x)-\langle \zeta, h\rangle}{|h|} < 0$, contradicting our assumption. This proves that $f(y) \ge f(x) + \langle \zeta, y - x\rangle$ for every $y \in X$, and therefore $\zeta \in \partial f(x)$.

Once established the equivalence between the five first assertions, we will prove that (i), (vi), and (vii) are equivalent too.

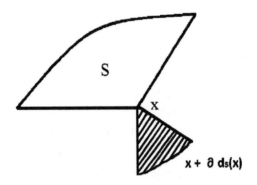

Fig. 3.2 Subdifferential of the distance function.

(i) $\Rightarrow$ (vi). It is trivial since the function $\varphi(y) = f(x) + \langle \zeta, y - x \rangle$ satisfies the required conditions.

(vi) $\Rightarrow$ (vii). $\varphi(y) = f(x) + \langle \zeta, y - x \rangle + D^2\varphi(z_y)(y - x, y - x)$ where $z_y \in [x, y]$ for y close enough to x by Taylor's Formula. But $D^2\varphi$ is continuous, hence there exist $r > 0$ and $\sigma > 0$ such that $\|D^2\varphi(z_y)\| \leq \sigma$ provided that $z_y \in B(x, r)$, observing that this condition follows from $y \in B(x, r)$. We deduce that for $y \in B(x, r)$ we have:

$$f(y) \geq \varphi(y) = f(x) + \langle \zeta, y - x \rangle + D^2\varphi(z_y)(y - x, y - x)$$
$$\geq f(x) + \langle \zeta, y - x \rangle - \sigma |y - x|^2$$

(vii) $\Rightarrow$ (i). For $y \in B(x, r)$ and $t \in (0, 1]$, we have

$$(1-t)f(x) + tf(y) \geq f((1-t)x + ty) \geq f(x) + \langle \zeta, t(y-x) \rangle - \sigma |t(y-x)|^2$$

hence
$$t(f(y) - f(x)) \geq t\langle \zeta, y - x \rangle - t^2\sigma |y - x|^2$$

dividing by t and letting $t \downarrow 0$ we get $f(y) - f(x) \geq \langle \zeta, y - x \rangle$. $\square$

3.2 TWO EXAMPLES

Let S be a convex closed subset of $\mathbf{R}^n$, we know that the distance function d_S is convex. The aim of the following example is to study $\partial d_S(x)$ (see Fig. 3.2).

Example.   Let us observe first that if $x \in int\,S$, then $\partial d_S(x) = \{\nabla d_S(x)\} = \{0\}$. On the other hand, as we have seen before, $d_S^2(x) = (\delta_S)_{\frac{1}{2}}$

hence d_S^2 is continuously differentiable, and $\nabla d_S^2(x) = 2(x - x_0)$ where $x_0 \in S$ is the unique closest point to x, by Theorem 2.23. Consequently if $x \notin S$, we have that d_S is differentiable at x and $\nabla d_S(x) = \frac{1}{|x-x_0|}(x - x_0)$ by the Chain Rule. What happens when $x \in \partial S$? It is easy to see that, in general we do not have differentiability. Consider for instance the case $S = \{0\}$, the distance function d_S is the absolute value which is not differentiable at 0. We proceed to develop this specific case.

Let $x \in \partial S$. If $\zeta \in \partial f(x)$, and consequently $d_S(y) \geq \langle \zeta, y - x \rangle$, we have that $\langle \zeta, y - x \rangle \leq 0$ for every $y \in S$, hence ζ is a support of S at x, but also $|y - x| \geq d_S(y) \geq \langle \zeta, y - x \rangle$ holds for every $y \in \mathbf{R}^n$, which implies $|\zeta| \leq 1$. Therefore

$$\partial d_S(x) \subset \{\zeta \in \overline{B}(x, 1) : \langle \zeta, y - x \rangle \leq 0 \text{ for every } y \in S\}$$

Conversely, if $\langle \zeta, y - x \rangle \leq 0$ for every $y \in S$, in order to prove that $\zeta \in \partial d_S(x)$, we only need to check that $d_S(y) \geq \langle \zeta, y - x \rangle$ whenever $\langle \zeta, y - x \rangle > 0$. Let us denote the hyperplane $\{y : \langle \zeta, y - x \rangle = 0\}$ by H. If y satisfies $\langle \zeta, y - x \rangle > 0$, then $d_S(y) \geq d_H(y)$, but it is an exercise to prove that $d_H(y) = \langle \frac{1}{|\zeta|}\zeta, y - x \rangle \geq \langle \zeta, y - x \rangle$ since $|\zeta| \leq 1$. We have finished the characterization of $\partial d_S(x)$ as the set

$$\{\zeta \in \overline{B}(x, 1) : \langle \zeta, y - x \rangle \leq 0 \text{ for every } y \in S\}. \quad \text{(See Figure 3.2)}$$

Now we present another example of the subdifferential of a convex function.

Example. We consider the function $f : \mathbf{R}^n \to \mathbf{R}$ defined as

$$f(x_1, \ldots, x_n) = \max_{i=1,\ldots,n} x_1$$

First of all we observe that f is convex since

$$f(tx + (1 - t)y) = \max_{i=1,\ldots,n} (tx_i + (1 - t)y_i)$$

$$\leq t \max_{i=1,\ldots,n} x_i + (1 - t) \max_{i=1,\ldots,n} y_i = tf(x) + tf(y)$$

In order to characterize $\partial f(x)$ we must find the vectors ζ such that $f(y) \geq f(x) + \langle \zeta, y - x \rangle$. When $y = x + te_i$ this condition reads as $f(y) \geq f(x) + t\langle \zeta, e_i \rangle = f(x) + t\zeta_i$. If $t < 0$, $f(x) \geq f(y) \geq f(x) + t\zeta_i$ hence

$\zeta_i \geq 0$. On the other hand, if $x_i < f(x)$ we have that $f(y) = f(x)$ and consequently $0 \leq t\zeta_i$ for $|t|$ small enough, which implies $\zeta_i = 0$. Let us denote by I the set of indices i such that $f(x) = x_i$, particularizing the formula with $y = x + t\sum_{i\in I} e_i, t > 0$, we get

$$f(x) + t = f(y) \geq f(x) + t\left\langle \zeta, \sum_{i\in I} e_i \right\rangle = f(x) + t\sum_{i\in I}\zeta_i$$

which implies $\sum_{i\in I}\zeta_i \leq 1$, while if $t < 0$, small,

$$f(x) - t = f(y) \geq f(x) + t\left\langle \zeta, \sum_{i\in I} e_i \right\rangle = f(x) + t\sum_{i\in I}\zeta_i$$

which implies $\sum_{i\in I}\zeta_i \geq 1$. Therefore

$$\partial f(x) \subset \left\{ \zeta : \sum_{i\in I}\zeta_i = 1, \ \zeta_i \geq 0, \quad \text{and} \quad \zeta_i = 0 \text{ for } i \notin I \right\}.$$

We are going to see that actually, this property characterizes the subgradients of f.

We consider a vector ζ with $\zeta_i = 0$ for $i \notin I$, $\zeta_i \geq 0$ always, and $\sum_{i=1}^n \zeta_i = 1$. We have to prove that

$$f(x + h) \geq f(x) + \sum_{i=1}^n h_i\zeta_i = f(x) + \sum_{i\in I}h_i\zeta_i. \qquad (3.2)$$

Let us observe that for $|h|$ small enough, $f(x + h) = f(x + \bar{h})$ where $\bar{h}_i$ is equal to h_i when $i \in I$ and $\bar{h}_i = 0$ otherwise. Moreover $f(x + \bar{h}) = f(x) + f(\bar{h})$ in this case, hence we only have to check that $\sum_{i\in I} h_i\zeta_i \leq f(\bar{h})$ which is immediate. Once established (3.2) for a small h, convexity gives us the formula for every h. Consequently, we have seen that

$$\partial f(x) = \left\{ \zeta : \sum_{i\in I}\zeta_i = 1, \ \zeta_i \geq 0, \quad \text{and} \quad \zeta_i = 0 \text{ for } i \notin I \right\}$$

3.3 PROBLEMS

(1) Consider the function $\delta_{[0,1]} : \mathbf{R} \to (-\infty, +\infty]$. Calculate its subdifferential at every point $x \in dom(\delta_{[0,1]}) = [0, 1]$.

(2) Give an example of a *lsc* convex function $f : \mathbf{R} \to (-\infty, +\infty]$ such that $\partial f(x) = \emptyset$ for a point $x \in \partial(\mathrm{dom}f)$.

(3) Give an example of a, necessarily nonconvex, C^1 function $f : \mathbf{R} \to \mathbf{R}$ such that there is not any C^2 function $g : \mathbf{R} \to \mathbf{R}, g \leq f$ in a neighborhood of 0, with $g(0) = f(0)$ and $g'(0) = f'(0)$. (Asserts (ii) and (vi) in Theorem 3.7 are not equivalent for nonconvex functions).

(4) Draw some level curves of the distance function d_S for the closed convex set $S = [-1, 1]^2 \subset \mathbf{R}^2$.

(5) Calculate the subdifferential of the following norms in $\mathbf{R}^n$:

$$||x||_1 = |x_1| + \cdots + |x_n| \quad \text{and} \quad ||x||_\infty = \max\{|x_1|, \ldots, |x_n|\}$$

(6) Prove that for a closed convex set S, the distance function d_S cannot be differentiable at the boundary points $x \in \partial S$.

(7) Characterize the points of differentiability of the function

$$f(x_1, \ldots, x_n) = \max_{i=1,\ldots,n} x_1$$

(8) Let us consider k *lsc* convex functions $f_i : X \to (-\infty, +\infty]$. Let us define $f(x) = \max_{i=1,\ldots,k} f_i(x)$. Prove that for a $x_0 \in \bigcap_{i=1,\ldots,k} \mathrm{dom} f_i$ we have

$$\partial f(x_0) = co\left(\bigcup_{i \in I(x_0)} \partial f_i(x_0) \right)$$

where $I(x_0) = \{i : f(x_0) = f_i(x_0)\}$.

CHAPTER 4

The Subdifferential: General Case

Convex functions are important for practical reasons as well as for their theoretical interest. The concept of subdifferential that we introduced in the preceding chapter allows us to study the minima of convex functions. It was presented in different equivalent versions, geometrical and analytical, and we saw it has nice topological and geometrical properties. And although we have not developed it yet—we are waiting to do so in a more general context—when working with the subdifferential, a practical and rich calculus holds. But what happens if we deal with a more general class of functions? With *lsc* functions for instance? It is clear that it is not possible to define the subdifferential in the same way as in the convex case. The definition that we gave is far too specific for that case. Even functions as nice as polynomials cannot satisfy that property, for example $f(t) = t^3$ at 0. The key for extending these ideas is Theorem 3.7. Assertions (ii) and (iii) are trivially satisfied by C^1 functions, with $\zeta = \nabla f(x)$, and for assertions (iv) and (v) we just need the function to be differentiable. We will not consider assertions (vi) and (vii) for the time being.

4.1 DEFINITION AND BASIC PROPERTIES

Before defining what we will call the subdifferential of a *lsc* function, we are going to establish some equivalences.

THEOREM 4.1. *Let $f : X \to (-\infty, +\infty]$ be a lsc function, $x \in dom f$ and $\zeta \in X$. The following assertions are equivalent:*

(i) *There exists a C^1 function $\varphi : X \to \mathbf{R}$ such that $\varphi(x) = f(x)$, $\nabla\varphi(x) = \zeta$, and $\varphi \leq f$ in a neighborhood of x.*

(ii) *There exists a differentiable function $\varphi : X \to \mathbf{R}$ such that $\varphi(x) = f(x)$, $\nabla\varphi(x) = \zeta$, and $\varphi \leq f$ in a neighborhood of x.*

(iii) $\liminf_{h \to 0} \dfrac{f(x+h) - f(x) - \langle \zeta, h \rangle}{|h|} \geq 0.$

An Introduction to Nonsmooth Analysis. http://dx.doi.org/10.1016/B978-0-12-800731-0.00004-7

PROOF. (i) $\Rightarrow$ (ii) is trivial. The proof of (ii) $\Rightarrow$ (iii) is identical to the convex case one. It only remains to prove (iii) $\Rightarrow$ (i). We claim that the result is true if $x = 0$, $\zeta = 0$, $f(x) = 0$, and f bounded below.

Assuming the claim, if we consider the function $g(h) = \max\{f(x + h) - f(x) - \langle \zeta, h \rangle, -1\}$, we have that $\liminf_{h \to 0} \frac{g(h)}{|h|} \geq 0$, and consequently there exists a C^1 function $\psi : X \to \mathbf{R}$ such that: $\psi(0) = 0 = g(0)$, $\nabla \psi(0) = 0$, and $\psi \leq g$ in a neighborhood of 0. The function $\varphi(y) = \psi(y - x) + \langle \zeta, y - x \rangle + f(x)$ is C^1, and satisfies $\nabla \varphi(x) = \zeta$, $\varphi(x) = f(x)$, and $\varphi \leq f$ in a neighborhood of x, since $g(y - x) = f(y) - f(x) - \langle \zeta, y - x \rangle$ in a neighborhood of x. Let us prove the claim.

We are assuming that $\liminf_{h \to 0} \frac{f(h)}{|h|} \geq 0$. Let us define $\rho : [0, +\infty) \to \mathbf{R}$ by $\rho(t) = \inf_{|h| \leq t} f(h)$, ρ is clearly nonincreasing and $\rho(0) = 0$, hence $\rho \leq 0$. On the other hand $\lim_{t \downarrow 0} \frac{\rho(t)}{t} = 0$. We define two new functions:

$$\rho_1(t) = \int_t^{et} \frac{\rho(s)}{s} ds \quad \rho_2(t) = \int_t^{et} \frac{\rho_1(s)}{s} ds$$

the first integral exists since $\frac{\rho(s)}{s}$ has a countable amount of discontinuities (ρ is monotone), moreover it defines a continuous function ρ_1. The second integral defines a C^1 function ρ_2. Both functions are nonincreasing, ρ_2 trivially since $\rho_2' \leq 0$, and ρ_1 because if $t_1 < t_2$ then

$$\rho_1(t_2) = \int_{t_2}^{et_2} \frac{\rho(s)}{s} ds = \int_{t_1}^{et_1} \frac{\rho\left(\frac{t_2}{t_1} r\right)}{\frac{t_2}{t_1} r} \frac{t_2}{t_1} dr$$

$$= \int_{t_1}^{et_1} \frac{\rho\left(\frac{t_2}{t_1} r\right)}{r} dr \leq \int_{t_1}^{et_1} \frac{\rho(r)}{r} ds = \rho_1(t_1)$$

where the second identity follows from the change of variable $s = \frac{t_2}{t_1} r$.

We are going to prove the following chain of inequalities:

$$\rho(e^2 t) \leq \rho_1(et) \leq \rho_2(t) \leq \rho_1(t) \leq \rho(t) \tag{4.1}$$

(1) $\rho_1(et) = \int_{et}^{e^2 t} \frac{\rho(s)}{s} ds \geq \int_{et}^{e^2 t} \frac{\rho(e^2 t)}{s} ds = \rho(e^2 t) \log e = \rho(e^2 t)$.

(2) $\rho_2(t) = \int_t^{et} \frac{\rho_1(s)}{s} ds \geq \int_t^{et} \frac{\rho_1(et)}{s} ds = \rho_1(et)$.

(3) $\rho_2(t) = \int_t^{et} \frac{\rho_1(s)}{s} ds \leq \int_t^{et} \frac{\rho_1(t)}{s} ds = \rho_1(t)$.

(4) $\rho_1(t) = \int_t^{et} \frac{\rho(s)}{s} ds \leq \int_t^{et} \frac{\rho(t)}{s} ds = \rho(t)$.

Dividing by t in (4.1), from $\lim_{t \downarrow 0} \frac{\rho(t)}{t} = 0$ we deduce

$$\lim_{t \downarrow 0} \frac{\rho_2(t)}{t} = \lim_{t \downarrow 0} \frac{\rho_1(t)}{t} = \lim_{t \downarrow 0} \frac{\rho(t)}{t} = 0 \qquad (4.2)$$

We can now define the function φ as $\varphi(0) = 0$ and $\varphi(h) = \rho_2(|h|)$ otherwise. We have that $\nabla \varphi(0) = 0$ since $\lim_{h \to 0} \frac{\varphi(h)}{|h|} = 0$ and

$$\nabla \varphi(h) = \rho_2'(|h|) \frac{h}{|h|} = \left(\frac{\rho_1(e|h|)}{e|h|} - \frac{\rho_1(|h|)}{|h|} \right) \frac{h}{|h|} \qquad (4.3)$$

since $\nabla(|\ |)(h) = \frac{h}{|h|}$. In order to prove that φ is C^1, we only have to check that $\lim_{h \to 0} \nabla \varphi(h) = 0$. This follows from (4.2) and (4.3) since

$$\lim_{h \to 0} |\nabla \varphi(h)| \leq \lim_{t \downarrow 0} \left| \frac{\rho_1(et)}{et} \right| + \lim_{t \downarrow 0} \left| \frac{\rho_1(t)}{t} \right| = 0$$

Finally, $f(z) \geq \rho(|z|) \geq \rho_2(|z|) = \varphi(z)$. $\qquad \square$

Let us observe that the inequality $\varphi(y) \leq f(y)$ is strict if $y \neq x$, replacing the function φ by $\varphi(y) - |y - x|^2$ for instance. On the other hand, the existence of a global C^1 function that satisfies property (i) is also equivalent to the other asserts under very mild conditions, namely if f is bounded on bounded sets (in this case we would be able to define ρ), this happens when $X = \mathbf{R}^n$ for instance. Let us observe that the requirement of lower semi-continuity is redundant since condition (iii) implies it.

We proceed to define the subdifferential through these equivalences.

DEFINITION 4.2. For a *lsc* function $f : X \to (-\infty, +\infty]$, and a point $x \in dom f$, we define the subdifferential of f at x, as the set of all vectors $\zeta \in X$ that satisfy one of the equivalent conditions in Theorem 4.1. We will denote it by $\partial f(x)$.

It is clear that for convex functions both definitions agree. Many of the properties of the subdifferential of convex functions hold also in the general case.

PROPOSITION 4.3. *Let $f : X \to (-\infty, +\infty]$ be a lsc function and $x \in dom f$. The following properties hold:*

(i) If f attains a local minimum at x, then $0 \in \partial f(x)$.

(ii) $\partial f(x)$ is a convex closed subset of X.
(iii) If $x \in int(dom f)$ and f is differentiable at x, then $\partial f(x) = \{\nabla f(x)\}$.
(iv) If $x \in int(dom f)$ and f is Gâteaux differentiable at x, then $\partial f(x) \subset \{f'_G(x)\}$.

PROOF. For (i) it is enough to observe that $f(x + h) \geq f(x)$ for h small enough. Convexity in (ii) follows immediately from the definition through the lim inf. In order to see that $\partial f(x)$ is closed, we consider a sequence $\{\zeta_n\} \subset \partial f(x)$ such that $\lim \zeta_n = \zeta$. We have

$$\liminf_{h \to 0} \frac{f(x + h) - f(x) - \langle \zeta, h \rangle}{|h|}$$

$$= \liminf_{h \to 0} \left(\frac{f(x + h) - f(x) - \langle \zeta_n, h \rangle}{|h|} + \frac{\langle \zeta_n - \zeta, h \rangle}{|h|} \right)$$

$$\geq \liminf_{h \to 0} \frac{f(x + h) - f(x) - \langle \zeta_n, h \rangle}{|h|}$$

$$-|\zeta_n - \zeta| \geq -|\zeta_n - \zeta| \quad \text{for every } n.$$

This proves that $\liminf_{h \to 0} \frac{f(x+h)-f(x)-\langle \zeta, h \rangle}{|h|} \geq 0$ and therefore $\zeta \in \partial f(x)$.

Let us prove part (iv). If f is Gâteaux differentiable and $\zeta \in \partial f(x)$, we have that $\liminf_{h \to 0} \frac{f(x+h)-f(x)-\langle \zeta, h \rangle}{|h|} \geq 0$. Taking $h = tv, t > 0$, we get

$$\liminf_{t \downarrow 0} \frac{f(x + tv) - f(x) - t\langle \zeta, v \rangle}{t} \geq 0$$

and consequently

$$\langle \zeta, v \rangle \leq \liminf_{t \downarrow 0} \frac{f(x + tv) - f(x)}{t} = \langle f'_G(x), v \rangle.$$

We conclude that $\langle \zeta - f'_G(x), v \rangle \leq 0$ for every v, and therefore $f'_G(x) = \zeta$.

Finally, part (iii) is an immediate consequence of part (iv) ($\partial f(x) \subset \{\nabla f(x)\}$) and part (ii) of Theorem 4.1 ($\nabla f(x) \in \partial f(x)$). □

The inclusion in property (iv) is strict in general. The function $f : \mathbf{R}^2 \to \mathbf{R}$ defined by $f(x, y) = -x$ when $y = x^2, x > 0$, and $f(x, y) = 0$ otherwise, is *lsc*, Gâteaux differentiable at $(0,0)$, with $f'_G(0, 0) = (0, 0)$, but

$$\liminf_{(x,y)\to(0,0)} \frac{f(x,y) - f(0,0) - \langle(0,0),(x,y)\rangle}{|(x,y)|} = \liminf_{(x,y)\to(0,0)} \frac{f(x,y)}{|(x,y)|}$$

$$= \lim_{\delta\downarrow 0} \inf_{|(x,y)|<\delta} \frac{f(x,y)}{|(x,y)|} \leq \lim_{\delta\downarrow 0} \frac{-\frac{\sqrt{2}}{2}\delta}{\delta} = -\frac{\sqrt{2}}{2}$$

hence $\partial f(0,0) = \emptyset$.

This example proves also that the subdifferential of a *lsc* function can be empty. In other words, while a convex function is subdifferentiable everywhere, general functions may lack this property. However, as we will prove later, the set of points where a *lsc* function has a nonempty subdifferential is dense.

Although Lipschitz functions will be studied in Chapter 6, we present here an easy property of the subdifferential of a Lipschitz function.

PROPOSITION 4.4. *Let* $f : X \to (-\infty, +\infty]$ *be a lsc function,* $x_0 \in domf$. *If* f *is Lipschitz in a neighborhood of* x_0, *with constant* K, *then* $|\zeta| \leq K$ *for every* $\zeta \in \partial f(x_0)$.

PROOF. Let $\zeta \in \partial f(x_0)$. We take $h = t\zeta$ with $t > 0$ in the definition, and we have $\liminf_{t\to 0^+} \frac{f(x_0+t\zeta)-f(x_0)-t|\zeta|^2}{t|\zeta|} \geq 0$, hence

$$|\zeta| \leq \liminf_{t\to 0^+} \frac{f(x_0 + t\zeta) - f(x_0)}{t|\zeta|} \leq K \qquad \square$$

We consider now the opposite situation, and we introduce the superdifferential of an *usc* function.

DEFINITION 4.5. We say that an *usc* $f : X \to [-\infty, +\infty)$ is superdifferentiable at $x \in domf$ if there exists a vector $\zeta \in X$ such that

$$\limsup_{h\to 0} \frac{f(x+h) - f(x) - \langle\zeta, h\rangle}{|h|} \leq 0.$$

The set of all such ζ is denoted by $\partial^+ f(x)$ and is called the superdifferential of f at x.

It is a mere exercise to observe that f is superdifferentiable at x if and only if $-f$ is subdifferentiable at x, and in this case the following formula holds:

$$\partial^+ f(x) = -\partial^-(-f)(x)$$

We list some results on superdifferentiability whose proof is a translation of the corresponding results on subdifferentiability via the above formula.

PROPOSITION 4.6. *Let* $f : X \to [-\infty, +\infty)$ *be an usc function and* $x \in dom f$. *The following assertions are equivalent:*

(i) $\zeta \in \partial^+ f(x)$.
(ii) *There exists a* C^1 *function* $\varphi : X \to \mathbf{R}$ *such that* $\varphi(x) = f(x)$, $\nabla\varphi(x) = \zeta$, *and* $\varphi \geq f$ *in a neighborhood of* x.
(iii) *There exists a differentiable function* $\varphi : X \to \mathbf{R}$ *such that* $\varphi(x) = f(x)$, $\nabla\varphi(x) = \zeta$, *and* $\varphi \geq f$ *in a neighborhood of* x.

PROPOSITION 4.7. *Let* $f : X \to [-\infty, +\infty)$ *be an usc function and* $x \in dom f$. *The following properties hold:*

(i) *If* f *attains a local maximum at* x, *then* $0 \in \partial^+ f(x)$.
(ii) $\partial^+ f(x)$ *is a convex closed subset of* X.
(iii) *If* $x \in int(dom f)$ *and* f *is differentiable at* x, *then* $\partial^+ f(x) = \{\nabla f(x)\}$.
(iv) *If* $x \in int(dom f)$ *and* f *is Gâteaux differentiable at* x, *then* $\partial^+ f(x) \subset \{f'_G(x)\}$.

We have seen that a differentiable function is both super and subdifferentiable. As a matter of fact that is a characterization of differentiable functions.

PROPOSITION 4.8. *Let* $f : X \to \mathbf{R}$ *be a continuous function, and* $x \in X$. *Then* f *is differentiable at* x *if and only if* $\partial^+ f(x) \cap \partial^- f(x) \neq \emptyset$. *Moreover, in this case,* $\partial^+ f(x) \cap \partial^- f(x) = \{\nabla f(x)\}$.

PROOF. The only if is an immediate consequence of the already proved results. Conversely, if a vector $\zeta \in \partial^+ f(x) \cap \partial^- f(x)$, we have

$$\limsup_{h \to 0} \frac{f(x + h) - f(x) - \langle \zeta, h \rangle}{|h|} \leq 0$$

and

$$\liminf_{h \to 0} \frac{f(x + h) - f(x) - \langle \zeta, h \rangle}{|h|} \geq 0$$

joining both inequalities we get $\lim_{h \to 0} \frac{f(x+h)-f(x)-\langle\zeta,h\rangle}{|h|} = 0$ hence f is differentiable at x and $\zeta = \nabla f(x)$.   □

The condition $\partial^+ f(x) \cap \partial^- f(x) \neq \emptyset$ can be relaxed, in fact it is enough for both $\partial^+ f(x)$ and $\partial^- f(x)$ to be nonempty, since if $\limsup_{h \to 0} \frac{f(x+h)-f(x)-\langle \zeta_1, h \rangle}{|h|} \leq 0$ and $\liminf_{h \to 0} \frac{f(x+h)-f(x)-\langle \zeta_2, h \rangle}{|h|} \geq 0$, it is immediate to see that $\zeta_1 = \zeta_2$.

An interesting consequence of Proposition 4.8 is the following characterization of differentiability.

COROLLARY 4.9. *A function $f : X \to \mathbf{R}$ is differentiable at x if and only if there are two C^1 functions $\varphi, \psi : X \to \mathbf{R}$ such that: $\varphi(x) = f(x) = \psi(x)$, and $\varphi \leq f \leq \psi$ in a neighborhood of x. Moreover, in such a case, $\nabla f(x) = \nabla \varphi(x) = \nabla \psi(x)$.*

In the preceding chapter we developed an exhaustive study of the distance function to a convex set d_S when $X = \mathbf{R}^n$. However things are more difficult if S is a, not necessarily convex, closed set. We illustrate it in the following example.

Example. The distance function d_S: First part.

Let $S \subset X$ closed. If $x \in int\, S$, then d_S is identically 0 in a neighborhood of x and consequently d_S is differentiable at x with $\nabla d_S(x) = 0$. Let us suppose now that $x \notin S$. We are going to assume that there is a point $\bar{x} \in S$ such that $d_S(x) = |x - \bar{x}|$. This is always true if $X = \mathbf{R}^n$, but in Hilbert spaces in general we cannot guarantee this fact. We are going to prove that $\frac{x-\bar{x}}{|x-\bar{x}|} \in \partial^+ d_S(x)$. In order to see it we study the quotient

$$\frac{d_S(x+h) - d_S(x) - \langle \frac{x-\bar{x}}{|x-\bar{x}|}, h \rangle}{|h|} = \frac{d_S(x+h) - |x-\bar{x}| - \langle \frac{x-\bar{x}}{|x-\bar{x}|}, h \rangle}{|h|}.$$

We will write a chain of inequalities, and then we will take the upper limit when $h \to 0$.

$$\frac{d_S(x+h) - |x-\bar{x}| - \langle \frac{x-\bar{x}}{|x-\bar{x}|}, h \rangle}{|h|} \leq \frac{|x-\bar{x}+h| - |x-\bar{x}| - \langle \frac{x-\bar{x}}{|x-\bar{x}|}, h \rangle}{|h|}$$

$$= \frac{1}{|h|} \left(\frac{|h|^2 + 2\langle x-\bar{x}, h \rangle}{|x-\bar{x}+h| + |x-\bar{x}|} - \left\langle \frac{x-\bar{x}}{|x-\bar{x}|}, h \right\rangle \right)$$

$$= \frac{|h|}{|x-\bar{x}+h| + |x-\bar{x}|} + \left(\frac{2}{|x-\bar{x}+h| + |x-\bar{x}|} - \frac{1}{|x-\bar{x}|} \right)\left\langle x-\bar{x}, \frac{h}{|h|} \right\rangle$$

$$\leq \frac{|h|}{|x-\bar{x}+h| + |x-\bar{x}|} + \left| \frac{2}{|x-\bar{x}+h| + |x-\bar{x}|} - \frac{1}{|x-\bar{x}|} \right| |x-\bar{x}|.$$

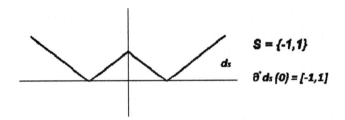

$S = \{-1,1\}$

$\partial^{\cdot} d_S (0) = [-1,1]$

Fig. 4.1 The superdifferential of the distance function.

This expression goes to 0 when $h \rightarrow 0$, hence $\limsup_{h \to 0}$ $\frac{d_S(x+h)-d_S(x)-(\frac{x-\bar{x}}{|x-\bar{x}|},h)}{|h|} \leq 0$. Therefore

$$\frac{x - \bar{x}}{|x - \bar{x}|} \in \partial^+ d_S(x).$$

This fact implies obviously that when the distance is attained at more than one point $\bar{x}$, the function cannot be differentiable since differentiability implies that the superdifferential is a singleton. So we restrict now to the situation in which $d_S(x) = |x - \bar{x}|$ for a single $\bar{x} \in S$. (See Fig. 4.1)

To be continued.

4.2 GEOMETRICAL MEANING OF THE SUBDIFFERENTIAL

Before studying the subdifferential of the distance function, we are going to introduce a new concept, as well as a useful Lemma.

DEFINITION 4.10. For a function $f : X \rightarrow (-\infty, +\infty]$ and a point $x \in domf$, we define the subderivative function $df(x) : X \rightarrow [-\infty, +\infty]$ by

$$df(x)(w_0) = \liminf_{t \downarrow 0, w \to w_0} \frac{f(x + tw) - f(x)}{t}$$

The lower limit of above means precisely

$$\lim_n \left[\inf_{0 < t < \frac{1}{n}, w \in B(w_0, \frac{1}{n})} \frac{f(x + tw) - f(x)}{t} \right].$$

It is an exercise to see that the subderivative is a positively homogeneous function. On the other hand the notation is clearly consistent since for a differentiable function the subderivative function is the differential.

LEMMA 4.11. *Let* $f : \mathbf{R}^n \to (-\infty, +\infty]$ *be a lsc function and* $x \in$ *dom*f *a point. We have*

$$\partial f(x) = \{\zeta : \langle \zeta, w_0 \rangle \leq df(x)(w_0) \quad \text{for every } w_0 \in X\}$$

PROOF. If $\zeta \in \partial f(x)$, we have $\liminf_{h \to 0} \frac{f(x+h)-f(x)-\langle \zeta, h \rangle}{|h|} \geq 0$, for an arbitrary nonzero $w_0 \in X$, taking $h = tw$, we have

$$\liminf_{t \downarrow 0, w \to w_0} \frac{f(x + tw) - f(x) - \langle \zeta, tw \rangle}{t|w|} \geq 0,$$

equivalently

$$\liminf_{t \downarrow 0, w \to w_0} \frac{f(x + tw) - f(x) - \langle \zeta, tw \rangle}{t} \geq 0,$$

which implies

$$df(x)(w_0) = \liminf_{t \downarrow 0, w \to w_0} \frac{f(x + tw) - f(x)}{t} \geq \lim_{w \to w_0} \langle \zeta, w \rangle = \langle \zeta, w_0 \rangle.$$

Conversely, let us assume that $\langle \zeta, w_0 \rangle \leq df(x)(w_0)$ for every $w_0 \in X$, if $\zeta \notin \partial f(x)$, from the definition we deduce that there exists a sequence $\{h_n\}$ such that $\lim h_n = 0$ and $\lim_n \frac{f(x+h_n)-f(x)-\langle \zeta, h_n \rangle}{|h_n|} < 0$. If we denote $w_n = \frac{h_n}{|h_n|}$ and $t_n = |h_n|$, the inequality above reads $\lim_n \frac{f(x+t_n w_n)-f(x)-\langle \zeta, t_n w_n \rangle}{t_n} < 0$. We may assume that $\lim_n w_n = w_0$ by the unit sphere compactness, hence

$$df(x)(w_0) = \liminf_{t \downarrow 0, w \to w_0} \frac{f(x + tw) - f(x)}{t}$$
$$\leq \lim_n \frac{f(x + t_n w_n) - f(x)}{t_n} < \lim_n \langle \zeta, w_n \rangle = \langle \zeta, w_0 \rangle,$$

which contradicts our assumption. $\qquad\qquad\qquad\qquad\qquad\square$

We continue with the study of the distance function.

Example. The distance function: Second Part.

As we announced in the first part, for a $x \notin S$, we only have to study $\partial d_S(x)$ when there is a unique $\bar{x}$ such that $d_S(x) = |x - \bar{x}|$. Our first goal is to prove that

$$d(d_S)(x)(w_0) = \frac{\langle x - \bar{x}, w_0 \rangle}{|x - \bar{x}|}$$

The equality holds trivially if $w_0 = 0$, consequently we may assume that $w_0 \neq 0$. It is clear that

$$\liminf_{t \downarrow 0, w \to w_0} \frac{d_S(x + tw) - d_S(x)}{t} = \liminf_{t \downarrow 0, w \to w_0} \frac{d_S(x + tw) - |x - \bar{x}|}{t}$$

$$\leq \liminf_{t \downarrow 0, w \to w_0} \frac{|x + tw - \bar{x}| - |x - \bar{x}|}{t}$$

but the norm is differentiable at $x - \bar{x} \neq 0$ with gradient $\frac{x - \bar{x}}{|x - \bar{x}|}$, hence

$$\liminf_{t \downarrow 0, w \to w_0} \frac{|x + tw - \bar{x}| - |x - \bar{x}|}{t}$$

$$= \liminf_{t \downarrow 0, w \to w_0} \frac{\left\langle \frac{x - \bar{x}}{|x - \bar{x}|}, tw \right\rangle + o(t|w|)}{t} = \left\langle \frac{x - \bar{x}}{|x - \bar{x}|}, w_0 \right\rangle.$$

For the opposite inequality we start with sequences $\{w_n\}$ converging to w_0 and $\{t_n\} \downarrow 0$, such that

$$\liminf_{t \downarrow 0, w \to w_0} \frac{d_S(x + tw) - d_S(x)}{t} = \lim_n \frac{d_S(x + t_n w_n) - d_S(x)}{t_n},$$

and selecting points $x_n \in S$ such that $d_S(x + t_n w_n) = |x + t_n w_n - x_n|$, we obtain a bounded sequence which, by compactness of closed bounded subsets of $\mathbf{R}^n$, we may suppose converges to a point that necessarily is $\bar{x}$. We have

$$\lim_n \frac{d_S(x + t_n w_n) - d_S(x)}{t_n}$$

$$= \lim_n \frac{|x + t_n w_n - x_n| - d_S(x)}{t_n} \geq \lim_n \frac{|x + t_n w_n - x_n| - |x - x_n|}{t_n}$$

$$\geq \lim_n \frac{\langle x - x_n, w_n \rangle}{|x - x_n|} = \frac{\langle x - \bar{x}, w_0 \rangle}{|x - \bar{x}|},$$

where the second inequality follows from the fact that the norm is convex and differentiable.

Once we have characterized the subderivative, we may apply Lemma 4.11 and we have that $\zeta \in \partial(d_S)(x)$ if and only if $\langle \zeta, w_0 \rangle \leq d(d_S)(x)(w_0) = \frac{\langle x - \bar{x}, w_0 \rangle}{|x - \bar{x}|}$ for every $w_0 \in \mathbf{R}^n$, which holds if and only if $\zeta = \frac{x - \bar{x}}{|x - \bar{x}|}$.

Joining this characterization with the results on superdifferentials, we have that in fact the distance function is differentiable at every point x such that the distance is attained at a single point $\bar{x}$, and $\nabla d_S(x) = \frac{x - \bar{x}}{|x - \bar{x}|}$.

When we studied the subdifferential of a convex function, the geometrical meaning was clear: there was a correspondence between the subdifferentials of the function and the supports of its epigraph. Things are more complicated now, as in general $epi\, f$ is not convex, and there are no supports. In order to give a geometrical translation of the subdifferential, we require some more tools.

DEFINITION 4.12. For a set $S \subset X$ and $x_0 \in S$, a vector $v \in X$ is normal to S at x_0 in the regular sense, written as $v \in \hat{N}_S(x_0)$, if

$$\langle v, x - x_0 \rangle \le o(|x - x_0|) \quad \text{for every} \quad x \in S \tag{4.4}$$

The set of all such vectors, $\hat{N}_S(x_0)$, is called the regular normal cone.

Let us observe that $0 \in \hat{N}_S(x_0)$ for every $x_0 \in S$, and that $\hat{N}_S(x_0) = \{0\}$ whenever $x \in int\, S$. Moreover $\hat{N}_S(x_0)$ is trivially a cone, that is: if $v \in \hat{N}_S(x_0)$ then $tv \in \hat{N}_S(x_0)$ for every $t \ge 0$.

It is not difficult to realize that formula (4.4) is another way of writing

$$\limsup_{x \to x_0, x \in S} \frac{1}{|x - x_0|} \langle v, x - x_0 \rangle \le 0. \tag{4.5}$$

(4.5) follows immediately from (4.4). On the other hand, if (4.5) holds, the function

$$\varphi(t) = t \sup \left\{ \left\langle v, \frac{z - x_0}{|z - x_0|} \right\rangle : z \in S \cap \overline{B}(x_0, t) \right\}$$

satisfies $\lim_{t \downarrow 0} \frac{\varphi(t)}{t} = 0$ and $\langle v, x - x_0 \rangle \le \varphi(|x - x_0|)$ for every $x \in S$, hence $\langle v, x - x_0 \rangle \le o(|x - x_0|)$ for every $x \in S$.

Another way of looking at $\hat{N}_S(x_0)$ is starting with tangent cones.

DEFINITION 4.13. A vector w is tangent to a set S at a point $x_0 \in S$, if there are sequences $\{x_n\} \in S$ and $t_n \downarrow 0$ such that

$$\lim_n \frac{x_n - x_0}{t_n} = w.$$

We call the set of all such vectors the tangent cone to S at x_0, and we denote it by $T_S(x_0)$.

It is immediate to see that the tangent cone is a cone, and that $T_S(x_0) = X$ if $x_0 \in int\, S$, but not only in this case.

PROPOSITION 4.14. $T_S(x_0)$ *is a closed cone.*

PROOF. Let $\{w_n\} \subset T_S(x_0)$ be a convergent sequence and let $\lim_n w_n = w$. For every k, we fix w_{n_k} such that $|w - w_{n_k}| < \frac{1}{k}$. For every k we choose a point $x_k \in S \cap B\left(x_0, \frac{1}{k}\right)$, and a real number $t_k \in \left(0, \frac{1}{k}\right)$ such that

$$\left| w_{n_k} - \frac{x_k - x_0}{t_k} \right| < \frac{1}{k}$$

It is clear that $\lim_k x_k = x_0$, $t_k \downarrow 0$ and $\lim_k \frac{x_k - x_0}{t_k} = w$, hence $w \in T_S(x_0)$. $\square$

The following formula links both concepts.

PROPOSITION 4.15. *Let* $X = \mathbf{R}^n$. $v \in \hat{N}_S(x_0)$ *if and only if* $\langle v, w \rangle \le 0$ *for every* $w \in T_S(x_0)$.

PROOF. Let $v \in \hat{N}_S(x_0)$ and $w \in T_S(x_0)$, $w = \lim \frac{x_n - x_0}{t_n}$ with $x_n \in S$, $\lim_n x_n = x_0$, and $t_n \downarrow 0$. We have

$$\langle v, w \rangle = \left\langle v, \lim \frac{x_n - x_0}{t_n} \right\rangle = \lim \frac{1}{t_n} \langle v, x_n - x_0 \rangle$$

$$\le \lim \sup \frac{o(|x_n - x_0|)}{t_n} = \lim \sup \frac{o(|x_n - x_0|)|x_n - x_0|}{t_n |x_n - x_0|} = 0.$$

Conversely, let us assume that $\langle v, w \rangle \le 0$ for every $w \in T_S(x_0)$. If $v \notin \hat{N}_S(x_0)$, from (4.5), we deduce that there is a sequence $\{x_n\} \subset S$, converging to x_0, such that $\lim \frac{1}{|x_n - x_0|} \langle v, |x_n - x| \rangle > 0$. We may assume, by passing to a subsequence if necessary, that the sequence $\left\{ \frac{x_n - x_0}{|x_n - x_0|} \right\}$ converges to a vector w. It is clear, taking $t_n = |x_n - x|$, that $w \in T_S(x_0)$ while $\langle v, w \rangle > 0$, which contradicts the assumption. $\square$

The next Corollary is a straightforward consequence of the fact that $\hat{N}_S(x_0)$ is the intersection of a family of closed half-spaces (see Fig. 4.2).

COROLLARY 4.16. *Let* $X = \mathbf{R}^n$. *Then* $\hat{N}_S(x_0)$ *is a closed convex set.*

A natural question arises here: is the tangent cone also convex? Clearly the answer is no. Consider for instance the set $S = \{(x, 0) \in \mathbf{R}^2 : x \le 0\} \cup \{(x, x) \in \mathbf{R}^2 : x \ge 0\}$ (See Fig. 4.2).

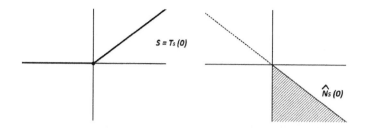

Fig. 4.2 Tangent cone and regular normal cone of S.

With these ingredients, we are able to characterize geometrically the subdifferential of a function.

THEOREM 4.17. *Let $f : \mathbf{R}^n \to (-\infty, +\infty]$ be a lsc function, $x_0 \in \text{dom} f$, then*

$$\partial f(x_0) = \{\zeta : (\zeta, -1) \in \hat{N}_{epif}((x_0, f(x_0))\} \quad \text{(See Fig. 4.3)}.$$

PROOF. By Lemma 4.11, a vector $\zeta \in \partial f(x_0)$ satisfies that $\langle \zeta, w_0 \rangle \leq r$ for every w_0 and every $r \geq df(x_0)(w_0)$. Another way of writing this is: $\langle (\zeta, -1), (w_0, r) \rangle = \langle \zeta, w_0 \rangle - r \leq 0$ for every $(w_0, r) \in epidf(x_0)$.

We claim that $epidf(x_0) = T_{epif}(x_0, f(x_0))$. This implies that $\zeta \in \partial f(x_0)$ if and only if $(\zeta, -1) \in \hat{N}_{epif}((x_0, f(x_0)))$. In order to conclude we must prove the claim.

From the definitions we have $(w_0, r) \in epidf(x_0)$ if and only if

$$\liminf_{t \downarrow 0, w \to w_0} \frac{f(x_0 + tw) - f(x_0)}{t} \leq r.$$

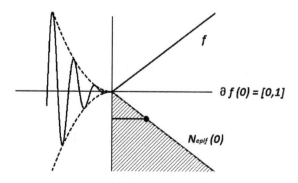

Fig. 4.3 Theorem 4.17.: example.

This inequality is equivalent to the existence of sequences $\{w_n\}$ converging to w_0 and $t_n \downarrow 0$ such that

$$\lim_n \frac{f(x_0 + t_n w_n) - f(x_0)}{t_n} \leq r. \tag{4.6}$$

On the other hand, $(w_0, r) \in T_{epif}(x_0, f(x_0))$ means that there are sequences $\{(x_n, \alpha_n)\}$ and $t_n \downarrow 0$, such that

$$\lim_n x_n = x_0, \quad \lim_n \alpha_n = f(x_0), \quad f(x_n) \leq \alpha_n,$$

$$w_0 = \lim_n \frac{x_n - x_0}{t_n}, \quad \text{and} \quad r = \lim_n \frac{\alpha_n - f(x_0)}{t_n}. \tag{4.7}$$

From (4.6), denoting $x_n = x_0 + t_n w_n$ and $\alpha_n = f(x_n) + t_n \left(r - \lim_n \frac{f(x_0 + t_n w_n) - f(x_0)}{t_n} \right)$, in order to verify (4.7) we only have to check that $\lim_n \alpha_n = \lim_n f(x_n) = f(x_0)$, which is true since

$$0 \leq \liminf f(x_n) - f(x_0) \leq \limsup f(x_n) - f(x_0) \leq 0,$$

where the first inequality follows by lower semi-continuity of f, while the second one follows from $\lim_n \frac{f(x_0 + t_n w_n) - f(x_0)}{t_n} \leq r$. The other conditions are trivially satisfied.

Starting with (4.7), we denote $w_n = \frac{x_n - x_0}{t_n}$, and it is clear that $\lim_n w_n = w_0$, and that

$$\lim_n \frac{f(x_0 + t_n w_n) - f(x_0)}{t_n} = \lim_n \frac{f(x_n) - f(x_0)}{t_n} \leq \lim_n \frac{\alpha_n - f(x_0)}{t_n} = r.$$

This proves (4.6). □

We will return one last time to the distance function.

Example. The distance function: Conclusion.

It remains to study the case $x \in \partial S$, when $S \subset \mathbf{R}^n$. In order to characterize it, we have to study the set $epi\, d_S$ at points $(x, d_S(x)) = (x, 0)$ with $x \in \partial S$, specifically the normal regular cone to $epi\, d_S$ at such points, consequently we start with the tangent cone. A point $(w, r) \in \mathbf{R}^{n+1}$ belongs to $T_{epid_S}(x, 0)$ if and only if there are sequences $\{x_n\}$ converging to x, $\{s_n\}$ converging to 0 and $t_n \downarrow 0$ such that

$$d_S(x_n) \leq s_n, \quad \lim_n \frac{x_n - x}{t_n} = w \quad \text{and} \quad \lim_n \frac{s_n}{t_n} = r.$$

It is trivial to observe that r must be greater than or equal to 0 since $s_n, t_n \geq 0$. When $w \in T_S(x)$, we may take the sequence $\{x_n\} \subset S$, and $s_n = rt_n$ with $r \in [0, +\infty)$ arbitrary, hence

$$\{(w, r) \in T_{epid_S}(x, 0) : w \in T_S(x)\} = T_S(x) \times [0, +\infty).$$

If $\zeta \notin \hat{N}_S(x)$, then there exists $w \in T_S(x)$ such that $\langle \zeta, w \rangle > 0$, as we have seen $(w, 0) \in T_{epid_S}(x, 0)$, but $\langle (\zeta, -1), (w, 0) \rangle = \langle \zeta, w \rangle > 0$. This implies $(\zeta, -1) \notin \hat{N}_{epid_S}(x, 0)$ and consequently $\zeta \notin \partial d_S(x)$.

If $|\zeta| > 1$ we have

$$\liminf_{h \to 0} \frac{d_S(x + h) - d_S(x) - \langle \zeta, h \rangle}{|h|}$$

$$\leq \liminf_{h \to 0} \frac{|h| - \langle \zeta, h \rangle}{|h|} \leq \liminf_{t \downarrow 0} \frac{|t\zeta| - t|\zeta|^2}{t|\zeta|} = 1 - |\zeta| < 0$$

hence $\zeta \notin \partial d_S(x)$. Joining both results, we have $\partial d_S(x) \subset \hat{N}_S(x) \cap \overline{B}(0, 1)$. The converse is also true, but we will prove it later.

With respect to the superdifferential, it is clear that $\partial^+ d_S(x) \subset \{0\}$ for every $x \in \partial S$ since

$$\limsup_{h \to 0} \frac{d_S(x + h) - \langle \zeta, h \rangle}{|h|}$$

$$\geq \limsup_{h \to 0} \frac{-\langle \zeta, h \rangle}{|h|} \geq \limsup_{t \downarrow 0} \frac{-\langle \zeta, -t\zeta \rangle}{t|\zeta|} = |\zeta| > 0$$

if $\zeta \neq 0$.

The set $S = \{(x, y) \in \mathbf{R}^2 : |y| \geq x^2\}$ provides us with an example of $\partial^+ d_S((0, 0)) = \{(0, 0)\}$, since

$$\limsup_{(x,y) \to (0,0)} \frac{d_S(x, y)}{|(x, y)|} \leq \limsup_{(x,y) \to (0,0)} \frac{x^2}{|(x, y)|} \leq \limsup_{(x,y) \to (0,0)} \frac{x^2}{|x|} = 0.$$

Our next theorem summarizes all the precedent results on differentiability of the distance function.

THEOREM 4.18. *Let $S \subset \mathbf{R}^n$ be a closed set. Then the distance function d_S has the following properties.*

(i) *If $x \in \text{int } S$ then d_S is differentiable at x and $\nabla d_S(x) = 0$.*

(ii) *If $x \notin S$ and $d_S(x)$ is attained at a single point $\bar{x} \in S$, then d_S is also differentiable and $\nabla d_S(x) = \frac{x - \bar{x}}{|x - \bar{x}|}$.*

(iii) *If $x \notin S$ but $d_S(x)$ is attained at more than one point, then $\partial^+ d_S(x) \supset \left\{ \frac{x - \bar{x}}{|x - \bar{x}|} : d_S(x) = |\bar{x} - x| \right\}$ and $\partial^- d_S(x) = \emptyset$.*

(iv) *If $x \in \partial S$ then $\partial^- d_S(x) = \hat{N}_S(x) \cap \overline{B}(0, 1)$. In particular if $\hat{N}_S(x) \neq \{0\}$ we have $\partial^+ d_S(x) = \emptyset$.*

PROOF. We only have to prove part (iv). In order to see it we are going to calculate the subderivative function, specifically we will prove the following formula:

$$d(d_S)(x)(w_0) = d(w_0, T_S(x)) \tag{4.8}$$

From the definition of the subderivative function, we have that

$$
\begin{aligned}
d(d_S)(x)(w_0) &= \liminf_{t \downarrow 0, w \to w_0} \frac{d_S(x + tw)}{t} = \liminf_{t \downarrow 0, w \to w_0} d\left(w, \frac{-x + S}{t} \right) \\
&\leq \liminf_{t \downarrow 0, w \to w_0} \left[d\left(w_0, \frac{-x + S}{t} \right) + |w - w_0| \right] \\
&= \liminf_{t \downarrow 0} d\left(w_0, \frac{-x + S}{t} \right).
\end{aligned}
$$

Let us see that

$$\liminf_{t \downarrow 0} d\left(w_0, \frac{-x + S}{t} \right) = d(w_0, T_S(x)). \tag{4.9}$$

If $t_n \downarrow 0$ is a sequence such that $\lim_n d\left(w_0, \frac{-x+S}{t_n} \right)$ exists, we consider a sequence $x_n \in S$ satisfying $d\left(w_0, \frac{-x+S}{t_n} \right) = |w_0 - \frac{x_n - x}{t_n}|$. We may assume, by compactness, that the sequence $\left\{ \frac{x_n - x}{t_n} \right\}$ converges, and we necessarily have $\lim_n \frac{x_n - x}{t_n} \in T_S(x)$. Hence $\lim_n d\left(w_0, \frac{-x+S}{t_n} \right) \geq d(w_0, T_S(x))$. Consequently, we have

$$\liminf_{t \downarrow 0} d\left(w_0, \frac{-x + S}{t} \right) \geq d(w_0, T_S(x))$$

Conversely, $d(w_0, T_S(x)) = |w_0 - \lim_n \frac{x_n - x}{t_n}| = \lim_n |w_0 - \frac{x_n - x}{t_n}|$ for specific sequences $\{x_n\} \subset S$, $\lim_n x_n = x$ and $t_n \downarrow 0$. Hence

$$\liminf_{t \downarrow 0} d\left(w_0, \frac{-x + S}{t}\right) \leq d(w_0, T_S(x))$$

Both inequalities together give us formula (4.9), and therefore (4.8) holds. Once established the formula, Lemma 4.11 allows us to characterize the subgradients $\zeta \in \partial^- d_S(x)$ as the vectors satisfying

$$\langle \zeta, w_0 \rangle \leq d(w_0, T_S(x)) \quad \text{for every} \quad w_0.$$

For every w_0 we denote by $\bar{w}_0 \in T_S(x)$ a point satisfying $|w_0 - \bar{w}_0| = d(w_0, T_S(x))$. If $\zeta \in \hat{N}_S(x) \cap \overline{B}(0, 1)$ then

$$\langle \zeta, w_0 \rangle = \langle \zeta, w_0 - \bar{w}_0 \rangle + \langle \zeta, \bar{w}_0 \rangle \leq \langle \zeta, w_0 - \bar{w}_0 \rangle$$
$$\leq |\zeta| \|w_0 - \bar{w}_0\| \leq |w_0 - \bar{w}_0| = d(w_0, T_S(x)),$$

which proves the inclusion $\hat{N}_S(x) \cap \overline{B}(0, 1) \subset \partial^- d_S(x)$. The other inclusion was proved in the preceding example. (See Fig. 4.4(a))   □

We have already seen that $\partial^+ d_S(x)$ can be nonempty (necessarily $\{0\}$ provided that $\hat{N}_S(x) = \{0\}$). As a matter of fact d_S is differentiable in that example. Nevertheless $\hat{N}_S(x) = \{0\}$ does not imply the differentiability of d_S (See Fig. 4.4(b)).

Example. Let $S = \{(x, y) \in \mathbf{R}^2 : y \leq |x|\}$, $(0, 0) \in \partial S$, $\hat{N}_S(0, 0) = \{(0, 0)\}$, $\partial^- d_S(0, 0) = \{0\}$, but $\partial^+ d_S(0, 0) = \emptyset$ since

$$\limsup_{(x,y) \to (0,0)} \frac{d_S(x, y)}{|(x, y)|} \geq \limsup_{y \to 0^+} \frac{d_S(0, y)}{|(0, y)|} = \limsup_{y \to 0^+} \frac{\frac{\sqrt{2}}{2} y}{y} = \frac{\sqrt{2}}{2} > 0$$

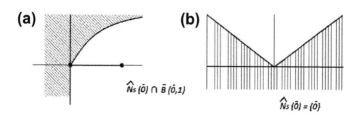

Fig. 4.4 Examples where the distance function is not differentiable.

Another interesting function is the indicator function and we are going to characterize its subdifferential.

Example. Let $S \subset \mathbf{R}^n$ be a closed set, we consider the function δ_S, clearly $dom\,\delta_S = S$. If $x_0 \in int\,S$, then δ_S is differentiable at x_0 and $\nabla \delta_S(x_0) = 0$. Our aim is to prove that

$$\partial \delta_S(x_0) = \hat{N}_S(x_0) \tag{4.10}$$

when $x_0 \in \partial S$. From the characterization of the regular normal cone that we gave in Proposition 4.15, it is enough to prove that $\zeta \in \partial \delta_S(x_0)$ if and only if $\langle \zeta, w \rangle \leq 0$ for every $w \in T_S(x_0)$. Taking this into account, by Lemma 4.11, $\zeta \in \partial \delta_S(x_0)$ if and only if $\langle \zeta, w \rangle \leq d(\delta_S)(x_0)(w)$ for every w. Let us first study the function $d(\delta_S)(x_0)$. We have

$$d(\delta_S)(x_0)(w) = \liminf_{t \downarrow 0, w' \to w} \frac{\delta_S(x_0 + tw') - \delta_S(x_0)}{t} = \liminf_{t \downarrow 0, w' \to w} \frac{\delta_S(x_0 + tw')}{t}$$

hence the range of $d(\delta_S)(x_0)$ is $\{0, +\infty\}$, and $d(\delta_S)(x_0)(w) = 0$ if and only if there are sequences $\{w_n\}$ converging to w and $t_n \searrow 0$ such that $x_0 + t_n w_n \in S$ for every n. This is clearly equivalent to $w \in T_S(x_0)$. Consequently, the assert $\langle \zeta, w \rangle \leq d(\delta_S)(x_0)(w)$ for every w is equivalent to $\langle \zeta, w \rangle \leq 0$ for every $w \in T_S(x_0)$.

4.3 DENSITY OF SUBDIFFERENTIABILITY POINTS

Although calculus will be the subject of the next chapter, we are going to present an elementary result of the subdifferential of the sum of two functions. First let us observe that the following formula is not true in general:

$$\partial(f + g)(x) = \partial f(x) + \partial g(x).$$

Consider for example functions $f, g : \mathbf{R} \to \mathbf{R}$ defined as $f(x) = |x|$, $g(x) = -|x|$. We have $(f + g)(x) = 0$ and consequently:

$$\partial(f + g)(0) = 0, \quad \partial f(0) = [-1, 1], \quad \partial g(0) = \emptyset$$

hence

$$\partial f(0) + \partial g(0) = [0, 1] + \emptyset = \emptyset \subset \{0\} = \partial(f + g)(0).$$

However we have the following result.

PROPOSITION 4.19. *Let $f : X \to (-\infty, +\infty]$ be a lsc function, $x \in$ dom$f$. Let $g : X \to \mathbf{R}$ be differentiable at x. Then*

$$\partial(f + g)(x) = \partial f(x) + \nabla g(x).$$

PROOF. Let $\nabla g(x) + \zeta \in \partial f(x) + \nabla g(x)$, then

$$\liminf_{h \to 0} \frac{(f + g)(x + h) - (f + g)(x) - \langle \nabla g(x) + \zeta, h \rangle}{|h|}$$

$$\geq \liminf_{h \to 0} \frac{f(x + h) - f(x) - \langle \zeta, h \rangle}{|h|}$$

$$+ \lim_{h \to 0} \frac{g(x + h) - g(x) - \langle \nabla g(x), h \rangle}{|h|}$$

$$= \liminf_{h \to 0} \frac{f(x + h) - f(x) - \langle \zeta, h \rangle}{|h|} \geq 0.$$

Conversely, if $\zeta \in \partial(f + g)(x)$, then there exists a function $\varphi : X \to \mathbf{R}$ differentiable at x, with $\nabla\varphi(x) = \zeta$, $(f+g)(x) = \varphi(x)$, and $\varphi \leq (f+g)$ in a neighborhood of x. The function $\psi = \varphi - g$ satisfies $\psi(x) = f(x)$, $\psi \leq f$ in a neighborhood of x and it is differentiable with differential $\nabla\psi(x) = \zeta - \nabla g(x)$, hence $\zeta - \nabla g(x) \in \partial f(x)$. $\square$

Let us observe that, with minor arrangements the first part of the proof works also in the general case. That is we have:

PROPOSITION 4.20. *Let $f, g : X \to (-\infty, +\infty]$ be two lsc functions, $x \in$ dom$f \cap$ dom$g$, then*

$$\partial f(x) + \partial g(x) \subset \partial(f + g)(x).$$

We will address now another important property of the subdifferential. As we have seen, there are points with an empty subdifferential, but however the set of points with nonempty subdifferential is dense. The result is true for Hilbert spaces, but we only prove the $\mathbf{R}^n$ version.

THEOREM 4.21. *Let $f : \mathbf{R}^n \to (-\infty, +\infty]$ be a lsc function. For every $x_0 \in$ dom$f$ and every $\varepsilon > 0$, there is a point x such that $|x - x_0| < \varepsilon$, $f(x_0) - \varepsilon \leq f(x) \leq f(x_0)$ and $\partial f(x) \neq \emptyset$. In particular, $\{x : \partial f(x) \neq \emptyset\}$ is dense in dom$f$.*

PROOF. From f being lsc we deduce that there is a positive δ, we may assume $\delta < \varepsilon$, such that $f(y) \geq f(x_0) - \varepsilon$ for every $y \in \overline{B}(x_0, \delta)$.

We define $g : \mathbf{R}^n \to (-\infty, +\infty]$ as $g(y) = (\delta^2 - |y - x_0|^2)^{-1}$ if $y \in B(x_0, \delta)$ and $g(y) = +\infty$ otherwise. The function g is lsc since $\lim_{|y-x_0| \to \delta} g(y) = +\infty$. Hence the function $f+g$ is lsc and consequently attains a minimum x on the compact $\overline{B}(x_0, \delta)$, $x \in B(x_0, \delta)$ necessarily. We have that $0 \in \partial(f + g)(x)$. On the other hand, as g is differentiable in $B(x_0, \delta)$, we have that $\partial(f + g)(x) = \partial f(x) + \nabla g(x)$, hence $-\nabla g(x) \in \partial f(x)$. Finally,

$$f(x) + \delta^{-2} \leq f(x) + g(x) = (f + g)(x) \leq (f + g)(x_0) = f(x_0) + \delta^{-2}$$

hence $f(x_0) - \varepsilon \leq f(x) \leq f(x_0)$. $\square$

COROLLARY 4.22. *For a closed set $S \subset \mathbf{R}^n$, the set of points x such that there is a unique $\bar{x} \in S$ satisfying $d_S(x) = |x - \bar{x}|$ is dense.*

PROOF. This set coincides with the set where d_S is subdifferentiable in $\mathbf{R}^n \setminus S$ plus S itself, hence it contains $\{x \in \mathbf{R}^n : \partial(d_S)(x) \neq \emptyset\}$, which is dense by Theorem 4.21. $\square$

Theorem 4.21 can be translated to the superdifferential of usc functions. Joining this with the fact that convex functions are everywhere subdifferentiable, we get:

COROLLARY 4.23. *Let $f : \mathbf{R}^n \to \mathbf{R}$ be a convex function. Then f is differentiable in a dense subset of $\mathbf{R}^n$.*

4.4 PROXIMAL SUBDIFFERENTIAL

We will finish this chapter by introducing another kind of subdifferential. At the beginning of the chapter we focused on Theorem 3.7, and studied some of its equivalences in the broader frame of not necessarily convex functions. We postponed the study of properties (vi) and (vii) since, as the next example shows, they are not equivalent to the other five.

Example. Consider the function $f : \mathbf{R} \to \mathbf{R}$ defined by $f(x) = -x^{4/3}$. It does not satisfy for $x = 0$ and any $\zeta \in \mathbf{R}$, property (vii), of Theorem 3.7. Let us prove it.

Assume on the contrary that there exist $\zeta \in \mathbf{R}, r > 0$, and $\sigma > 0$ such that $f(y) \geq f(0) + \langle \zeta, y \rangle - \sigma y^2$, equivalently $-y^{\frac{4}{3}} \geq y\zeta - \sigma y^2$ for every $y \in B(0, r)$. We have that $\zeta \leq \sigma y - y^{\frac{1}{3}}$ for $y > 0$ which implies $\zeta \leq 0$, and $\zeta \geq \sigma y - y^{\frac{1}{3}}$ for $y < 0$ which implies $\zeta \geq 0$. Hence $\zeta = 0$, but then $\sigma y^2 \geq y^{\frac{4}{3}}$ near 0, which is false. This contradiction proves that the function f, that has properties (iii), (iv), and (v) for $x = 0$ and $\zeta = 0$, does not satisfy property (vii). We will see that f does not satisfy property (vi) either, at $x = 0$ for any $\zeta \in \mathbf{R}$.

Before defining the new subdifferential, we will establish the equivalence between properties (vi) and (vii).

PROPOSITION 4.24. *Let $f : X \to (-\infty, +\infty]$ be a lsc function, $\zeta \in X$, and $x \in \mathrm{dom} f$, the following assertions are equivalent:*

(i) *There exists a C^2 function $\varphi : X \to \mathbf{R}$ such that $\varphi(x) = f(x)$, $\nabla\varphi(x) = \zeta$, and $\varphi \leq f$ in a neighborhood of x.*
(ii) *There exist $r > 0, \sigma > 0$ such that $f(y) \geq f(x) + \langle \zeta, y-x \rangle - \sigma|y-x|^2$ for every $y \in B(x, r)$.*

PROOF. (i) follows from (ii) trivially, for $\varphi(y) = f(x) + \langle \zeta, y - x \rangle - \sigma|y-x|^2$. Assume (i), Taylor's Formula for the C^2 function φ at x gives us

$$\varphi(y) = \varphi(x) + \langle \nabla\varphi(x), y - x \rangle + \frac{1}{2}\nabla^2\varphi(z)(y - x, y - x)$$

$$= f(x) + \langle \zeta, y - x \rangle + \frac{1}{2}\nabla^2\varphi(z)(y - x, y - x)$$

for y near to x and $z \in [x, y]$. The fact that φ is C^2 implies that $\frac{1}{2}|\nabla^2\varphi(z) (y-x, y-x)| \leq \sigma|x-y|^2$ for a positive σ near to x since $\nabla^2\varphi(z)$ depends continuously on z. Hence $\frac{1}{2}\nabla^2\varphi(z)(y - x, y - x) \geq -\sigma|x - y|^2$ and we have proved (ii). $\qquad\square$

Once we have established this equivalence, we proceed to define the proximal subdifferential.

DEFINITION 4.25. Let $f : X \to (-\infty, +\infty]$ be a lsc function and $x \in \mathrm{dom} f$. We say that a vector ζ is a proximal subgradient of f at x whenever it satisfies one of the equivalent conditions of Proposition 4.24. The set of all the proximal subgradients at x is called the proximal subdifferential of f at x and we will denote it by $\partial_P f(x)$.

The inclusion $\partial_P f(x) \subset \partial f(x)$ follows trivially from the fact that C^2 functions are C^1. The above example shows that we can have $\partial_P f(0) = \emptyset \subset \{0\} = \partial f(0)$. Of course we have $\partial_P f(x) = \partial f(x)$ for convex functions. All the properties that we summarize in the next proposition are either trivial or their proofs are deduced from those of the general case with minor arrangements.

PROPOSITION 4.26. *Let* $f : X \to (-\infty, +\infty]$ *be a lsc function and* $x \in dom f$, *the following properties hold:*

(i)   *If f attains a local minimum at x, then $0 \in \partial_P f(x)$.*
(ii) *$\partial_P f(x)$ is a convex subset of X.*
(iii) *If $x \in int(dom f)$ and f is C^2 in a neighborhood of x, then $\partial_P f(x) = \{\nabla f(x)\}$.*
(iv) *If $x \in int(dom f)$ and f is Gâteaux differentiable at x, then $\partial_P f(x) \subset \{f'_G(x)\}$.*
(v)   *If $g : X \to \mathbf{R}$ is a C^2 function in a neighborhood of x, then $\partial_P(f + g)(x) = \partial_P f(x) + \nabla g(x)$.*
(vi) *If $X = \mathbf{R}^n$, then $\{x \in dom f : \partial_P f(x) \neq \emptyset\}$ is dense in $dom f$.*

We have already seen that for a C^1 function we can have $\partial_P f(x) = \emptyset$. The next example proves that $\partial_P f(x)$ may fail to be closed.

Example. If we consider the one variable function defined by $f(t) = -t^{\frac{4}{3}}$ if $t < 0$ and $f(t) = t$ otherwise, it is not difficult to see that $\partial_P f(0) = (0, 1]$.

We may define also the proximal superdifferential for *usc* functions directly, but it is completely equivalent to defining it by $\partial^P f(x) = -\partial_P(-f)(x)$.

Proximal subgradients can be characterized in a geometrical way, via the epigraph, but we will not present this interesting theory since it requires the introduction of another kind of normal cone.

4.5 PROBLEMS

(1) Calculate the subdifferential of the following functions $f : \mathbf{R} \to \mathbf{R}$ at $t = 0$.

(a) $f(t) = t \sin \frac{1}{t}$ for $t \neq 0$, $f(0) = 0$.

(b) $f(a_n) = 0$ for every n, for a given sequence $\{a_n\}$ converging to 0, and $f(t) = |t|$ otherwise. (Hint: there are three different situations.)

(c)

$$f(t) = \begin{cases} -t & \text{if } t \leq 0, \\ t \sin \frac{1}{t} & \text{if } t > 0. \end{cases}$$

(d)

$$f(t) = \begin{cases} \sqrt{|t|} & \text{if } t \leq 0, \\ t \sin \frac{1}{t} & \text{if } t > 0. \end{cases}$$

(e)

$$f(t) = \begin{cases} -t & \text{if } t \leq 0, \\ t^{\frac{3}{2}} \sin \frac{1}{t} & \text{if } t > 0. \end{cases}$$

(2) Calculate $\partial f(0)$ for the following function.

$$f(t) = \begin{cases} -t^4 & \text{if } t \leq 0, \\ t^2 \sin \frac{1}{t} & \text{otherwise.} \end{cases}$$

Find a C^1 function satisfying condition (i) in Theorem 4.1.

(3) Prove that function $f : \mathbf{R} \to \mathbf{R}$ defined by

$$f(t) = \begin{cases} 0 & \text{if } t \leq 0, \\ \frac{1}{n(n-1)}[-(2n-1)t + 1] & \text{if } t \in \left[\frac{1}{n}, \frac{1}{n-1}\right], \\ +\infty & t > 1 \end{cases}$$

is differentiable at $t = 0$. Find a C^1 function g such that $g(0) = g'(0) = 0$ and $g \leq f$.

(4) Prove that for a continuous function $f : \mathbf{R}^n \to \mathbf{R}$, if both $\partial^- f(x)$ and $\partial^+ f(x)$ are not empty then f is differentiable at x.

(5) Let $S \subset \mathbf{R}$ be a nonconvex closed set, and $x \notin S$. Assume that there are two points $\bar{x}_1, \bar{x}_2 \in S$ such that $d_S(x) = |x - \bar{x}_1| = |x - \bar{x}_2|$. Prove that $\partial^+ f(x) = [-1, 1]$.

(6) Let $S = \{(x, y) \in \mathbf{R}^2 : \min\{x, y\} \leq 0\}$. Prove that

$$\partial^+ d_S(1, 1) = \{(u, v) : u + v = 1, u \geq 0, v \geq 0\}.$$

(7) Let $S = \{(x, y, z) : x^2 + y^2 + z^2 = 1\}$ be the unit sphere of $\mathbf{R}^3$. Prove that $\partial^+ d_S(\bar{0}) = \overline{B}(\bar{0}, 1)$.

(8) Give an example of a continuous function $f : \mathbf{R} \to \mathbf{R}$ such that $\partial^- f(0) = \partial^+ f(0) = \emptyset$.

(9) Let $f : \mathbf{R}^n \to (-\infty, +\infty]$ be a lsc function, $x_0 \in domf$ and $\zeta \in \partial f(x_0)$. Let us define the function $\varphi_i : \mathbf{R} \to (-\infty, +\infty]$ by $\varphi_i(t) = f(x_0 + te_i)$. Prove that $\zeta_i \in \partial \varphi_i(0)$ for every $i = 1, \ldots, n$ (Partial subgradients).

(10) Prove that the subderivative function $df(x)(w_0)$ is positively homogeneous.

(11) Calculate the subderivative function for the functions that appear in problems 1 and 2.

(12) Prove that if a function $f : X \to \mathbf{R}$ is differentiable at x then the subderivative at x satisfies $df(x)(w_0) = \langle \nabla f(x), w_0 \rangle$.

(13) Let $f : \mathbf{R}^2 \to \mathbf{R}$ be defined in polar coordinates by

$$f(\rho, \theta) = \begin{cases} \rho^2 |\tan \theta| & \text{if } \theta \in \left(-\frac{\pi}{2}, \frac{\pi}{2} \right), \\ \rho |\tan \theta| & \text{if } \theta \in \left(\frac{\pi}{2}, \frac{3\pi}{2} \right), \\ 0 & \text{otherwise.} \end{cases}$$

Prove that f is lsc. Calculate the subderivative function and the subdifferential at the origin.

(14) For $f : \mathbf{R}^2 \to \mathbf{R}$ defined by $f(x, y) = |x| - |y|$, calculate $\partial f(x_0, y_0)$ for every $(x_0, y_0) \in \mathbf{R}^2$.

(15) Calculate the tangent cone, $T_S(0, 0)$, and the normal regular cone, $\hat{N}_S(0, 0)$, at $(0, 0)$ for the following sets S:

(a) $S = \{(x, y) \in \mathbf{R}^2 : y \le x^2\}$.

(b) $S = \{(x, y) \in \mathbf{R}^2 : |y| \le x^{\frac{3}{2}}, x \ge 0\}$.

(c) $S = \{(x, y) \in \mathbf{R}^2 : x \le \sqrt{|y|}\}$.

(d) $S = \{(x, y) \in \mathbf{R}^2 : y \le |x|\}$.

(e) $S = \{(x, y) \in \mathbf{R}^2 : y = |x|\}$.

(16) Let $f : \mathbf{R}^n \to \mathbf{R}$ be a differentiable function in a neighborhood of a point x_0. Calculate the tangent and the regular normal cones to the graph of f at $(x_0, f(x_0))$.

(17) Observe that we may characterize $co(T_S(x_0))$ as the set of vectors w satisfying $\langle v, w \rangle \le 0$ for every $v \in \hat{N}_S(x_0)$.

(18) Prove that for two closed subsets of $\mathbf{R}^n$, S_1 and S_2, and a point $x \in S_1 \cap S_2$, we have

$$\hat{N}_{S_1}(x) + \hat{N}_{S_2}(x) \subset \hat{N}_{S_1 \cap S_2}(x)$$

but the equality does not hold in general.

(19) Let $f : \mathbf{R}^n \to (-\infty, +\infty]$ be a *lsc* function and let us denote its convexification by $\hat{f}$. Prove that if $f(x) = \hat{f}(x) \in \mathbf{R}$, then $\partial f(x) \neq \emptyset$.

(20) Consider the set $S = \{(x, y) \in \mathbf{R}^2 : x^2 = y^2\}$. Let $d_S : \mathbf{R}^2 \to \mathbf{R}$. We define function $\varphi_1 : \mathbf{R} \to \mathbf{R}$ by $\varphi_1(t) = d_S(t, 0)$. Prove that although, as we saw in Problem 9,

$$\pi_1\big(\partial d_S(0, 0)\big) \subset \partial \varphi_1(0),$$

the equality does not hold.

CHAPTER 5

Calculus

The aim of this chapter is to extend some calculus tools to the nonsmooth setting. We have already seen Fermat's Rule, and we have given a very restrictive Sum Rule too. Exact results concerning the sum of two functions are not true in general, as we saw above, therefore we will try to get approximate results.

5.1 SUM RULE

Let us start with the Sum Rule. As we saw in Proposition 4.19, an Exact Sum Rule, $\partial f(x) + \partial g(x) = \partial(f + g)(x)$, is true if one of the functions is differentiable. In the general case we only have the inclusion $\partial f(x) + \partial g(x) \subset \partial(f + g)(x)$. Given that the counterexample involved very nice functions (nondifferentiable of course), we do not expect a better Exact Sum Rule than the one already obtained. Consequently we focus on an approximate rule. We start with a Lemma

LEMMA 5.1. *Let* $f, g : X \to (-\infty, +\infty]$ *be two lsc functions, let* $C \subset X$ *be a compact convex set such that* $C \cap \mathrm{dom} f \cap \mathrm{dom} g \neq \emptyset$. *Then for any positive sequence* $\{r_n\} \nearrow +\infty$, *we have*

$$\lim_n \inf_{x,y \in C} \left(f(x) + g(y) + r_n |x - y|^2 \right) = \inf_{x \in C} \left(f(x) + g(x) \right).$$

PROOF. Let us observe first that the limit in the formula exists since the sequence is increasing. Let us consider the function $F_n(x, y) = f(x) + g(y) + r_n |x - y|^2$, this function attains a minimum over the compact $C \times C$ at a point $(x_n, y_n) \in C \times C$ since it is *lsc*. We may assume, passing to a subsequence if necessary that $\lim_n (x_n, y_n) = (x_0, y_0) \in C \times C$. If we fix a point $z \in C \cap \mathrm{dom} f \cap \mathrm{dom} g$, we have

$$m_f + m_g + r_n |x_n - y_n|^2 \leq f(x_n) + g(y_n) + r_n |x_n - y_n|^2 \leq f(z) + g(z),$$

with m_f and m_g lower bounds over the compact C, of the *lsc* functions f and g respectively. This implies that the sequence $\{r_n |x_n - y_n|^2\}$ is bounded, and therefore $\lim_n |x_n - y_n| = 0$ since $\lim_n r_n = +\infty$. We conclude that

An Introduction to Nonsmooth Analysis. http://dx.doi.org/10.1016/B978-0-12-800731-0.00005-9

$x_0 = y_0$, and

$$\inf_{x \in C} \left(f(x) + g(x) \right)$$
$$\leq f(x_0) + g(x_0) \leq \liminf_n f(x_n) + \liminf_n g(y_n)$$
$$\leq \liminf_n f(x_n) + \liminf_n g(y_n) + \liminf_n r_n |x_n - y_n|^2$$
$$\leq \liminf_n F_n(x_n, y_n) = \lim_n F_n(x_n, y_n)$$

since f and g are *lsc*. Hence $\inf_{x \in C}(f(x) + g(x)) \leq \lim_n F_n(x_n, y_n)$. This proves one of the inequalities, the other inequality is trivial since $F_n(x_n, y_n) \leq F_n(x, x) = f(x) + g(x)$ for every $x \in C$.

Let us observe that we have proved also that all the inequalities above are equalities, and $\inf_{x \in C}(f(x) + g(x))$ is attained at x_0.   □

This Lemma allows us to prove the following Fuzzy Sum Rule.

THEOREM 5.2. *Let $f, g : \mathbf{R}^n \to (-\infty, +\infty]$ be two lsc functions, let $z_0 \in \operatorname{dom} f \cap \operatorname{dom} g$. If $\zeta \in \partial(f + g)(z_0)$, then for every $\varepsilon > 0$ there are $\bar{x}, \bar{y} \in B(z_0, \varepsilon)$ such that $|f(z_0) - f(\bar{x})| < \varepsilon$, $|g(z_0) - g(\bar{y})| < \varepsilon$ and $\zeta \in \partial f(\bar{x}) + \partial g(\bar{y}) + \varepsilon B$.*

PROOF. Let $\varepsilon > 0$ and $\zeta \in \partial(f + g)(z_0)$. There exist a closed ball around z_0, C, with a radius smaller than ε, and a C^1 function φ such that $\nabla\varphi(z_0) = \zeta$ and such that the minimum of $f + g - \varphi$ over C is attained only at z_0, moreover $(f + g - \varphi)(z_0) = 0$ (Ref. Theorem 4.1). We may take C, reducing its radius if necessary, to be such that $|\nabla\varphi(y) - \nabla\varphi(z_0)| < \varepsilon$ for every $y \in C$. Let us consider now a sequence $\{r_n\} \nearrow +\infty$, and define

$$F_n(x, y) = f(x) + g(y) - \varphi(y) + r_n|x - y|^2$$

for $(x, y) \in C \times C$ and $+\infty$ otherwise.

The minimum of F_n over $C \times C$ is attained at a point (x_n, y_n), and applying Lemma 5.1 we have that $\lim_n F_n(x_n, y_n) = (f + g - \varphi)(z_0) = 0$. Considering F_n as a function of x and y alternatively, we have that

$$F_n(x, y_n) = f(x) + g(y_n) - \varphi(y_n) + r_n|x - y_n|^2$$

has a minimum at x_n, and

$$F_n(x_n, y) = f(x_n) + g(y) - \varphi(y) + r_n|x_n - y|^2$$

has a minimum at y_n.

As the functions φ and $|\ |^2$ are differentiable, we deduce that

$$0 \in \partial f(x_n) + 2r_n(x_n - y_n) \quad \text{and} \quad 0 \in \partial g(y_n) - \nabla\varphi(y_n) - 2r_n(x_n - y_n),$$

hence

$$-2r_n(x_n - y_n) \in \partial f(x_n) \quad \text{and} \quad \nabla\varphi(y_n) + 2r_n(x_n - y_n) \in \partial g(y_n).$$

Therefore $\nabla\varphi(y_n) \in \partial f(x_n) + \partial g(y_n)$, and consequently $\zeta = \nabla\varphi(z_0) \in \partial f(x_n) + \partial g(y_n) + \varepsilon B$.

Finally,

$$f(x_n) + g(y_n) - \varphi(y_n) + r_n|x_n - y_n|^2$$
$$= F_n(x_n, y_n) \le F_n(z_0, z_0) = f(z_0) + g(z_0) - \varphi(z_0),$$

which implies

$$f(x_n) + g(y_n) \le f(z_0) + g(z_0) + \varphi(y_n) - \varphi(z_0),$$

from where we deduce

$$\limsup_n (f(x_n) + g(y_n)) \le f(z_0) + g(z_0).$$

On the other hand, $f(z_0) \le \liminf_n f(x_n)$ and $g(z_0) \le \liminf_n g(y_n)$ since f and g are lsc and $\lim_n x_n = z_0 = \lim_n y_n$ by the lemma, hence $\lim_n(f(x_n) + g(y_n)) = f(z_0) + g(z_0)$. If $f(z_0) < \liminf_n f(x_n)$, we have arrived at the contradiction

$$f(z_0) + g(z_0) < \liminf_n f(x_n) + \liminf_n g(y_n)$$
$$\le \liminf_n (f(x_n) + g(y_n)) \le f(z_0) + g(z_0)$$

hence $f(z_0) = \liminf_n f(x_n)$ and, passing to a subsequence if necessary, $f(z_0) = \lim_n f(x_n)$ and consequently $g(z_0) = \lim_n g(y_n)$.

We now consider an index n_0 such that $|f(z_0) - f(\bar{x})| < \varepsilon$ and $|g(z_0) - g(\bar{y})| < \varepsilon$ with $\bar{x} = x_{n_0}$ and $\bar{y} = y_{n_0}$, and $\bar{x}, \bar{y}$ fulfill the required properties. $\qquad\square$

If as above we consider the functions $f(t) = |t|$ and $g(t) = -|t|$, we have that $0 \in \partial(f + g)(0)$, $\partial f(0) + \partial g(0) = \emptyset$, but we could take $\bar{y}$ as close to 0 as we like, and $0 \in \partial f(0) + \partial g(\bar{y})$. In this case we only require to shift point 0 a little in order to guarantee an Exact Sum Rule. In general the situation is not as nice.

This theorem encourages us to let $\varepsilon \to 0$, in order to get an exact formula. But what happens with $\lim_{\varepsilon \to 0} \partial f(\bar{x})$? We require a wider subdifferential.

DEFINITION 5.3. For a lsc function $f : \mathbf{R}^n \to (-\infty, +\infty]$ and a point $x_0 \in domf$, we define the limiting subdifferential of f at x_0, $\partial_L f(x_0)$, as the set of all vectors ζ such that there exist sequences $\{x_n\} \subset domf$ converging to x_0 and $\{\zeta_n\}$ converging to ζ such that $\zeta_n \in \partial f(x_n)$ and $\lim_n f(x_n) = f(x_0)$.

The condition $\lim_n f(x_n) = f(x_0)$ is fulfilled automatically if f is continuous at x_0. This concept can be defined in the wider setting of Hilbert spaces, but then we require only the weak convergence of the sequence $\{\zeta_n\}$. Of course we can define also the limiting superdifferential of an usc, $\partial^L f(x_0)$, and it is not difficult to see that

$$\partial^L f(x_0) = -\partial_L(-f)(x_0).$$

It is immediate from the definitions that

$$\partial_P f(x_0) \subset \partial f(x_0) \subset \partial_L f(x_0).$$

The second inclusion is not an equality in general, as can be seen in the following example.

Example. The real function $f(t) = -|t|$ satisfies that $\partial f(x_0) = \emptyset$, but $\partial_L f(x_0) = \{-1, 1\}$.

This example proves also that $\partial_L f(x_0)$ is not a convex set in general. However, it is easy to see, using a diagonal process, that $\partial_L f(x_0)$ is closed. Once defined the limiting subdifferential, we will try to prove an Exact Sum Rule.

PROPOSITION 5.4. *Let $f, g : \mathbf{R}^n \to (-\infty, +\infty]$ be two lsc functions, $z_0 \in domf \cap domg$, assume that one of the functions is Lipschitz in a neighborhood of z_0. Then*

$$\partial_L(f + g)(z_0) \subset \partial_L f(z_0) + \partial_L g(z_0).$$

PROOF. Let $\zeta \in \partial_L(f+g)(z_0)$, for every n we choose $z_n \in B(z_0, \frac{1}{n})$ and $\zeta_n \in \partial(f+g)(z_n)$ such that $|(f+g)(z_n)-(f+g)(z_0)| < \frac{1}{n}$ and $|\zeta_n-\zeta| < \frac{1}{n}$. Applying Theorem 5.2 for every n, we obtain points $x_n, y_n \in B(z_n, \frac{1}{n})$ and vectors $v_n \in \partial f(x_n)$, $w_n \in \partial g(y_n)$ such that

$$|\zeta_n - v_n - w_n| < \frac{1}{n} \quad |f(x_n) - f(z_n)| < \frac{1}{n} \quad |g(y_n) - g(z_n)| < \frac{1}{n}.$$

The fact that one of the functions, f or g is Lipschitz around z_0 implies that one of the sequences $\{v_n\}$ or $\{w_n\}$ is bounded, and therefore so is the other one. This allows us to assume, passing to a subsequence, that there exist $\lim_n v_n = v$ and $\lim_n w_n = w$. We have in particular that $\zeta = v + w$. In order to prove that $v \in \partial_L f(z_0)$ and $w \in \partial_L g(z_0)$, and consequently $\zeta \in \partial_L f(z_0) + \partial_L f(z_0)$, it only remains to prove that $\lim_n f(x_n) = f(z_0)$ and $\lim_n g(y_n) = g(z_0)$. It is clear that $\lim_n (f(x_n) + g(y_n)) = f(z_0) + g(z_0)$. We have also that $f(z_0) \leq \liminf_n f(x_n)$ and $g(z_0) \leq \liminf_n g(y_n)$ since f and g are lsc. If one of this inequalities is strict, we would have

$$f(z_0) + g(z_0) < \liminf_n f(x_n) + \liminf_n g(y_n) \leq \liminf_n (f(x_n) + g(y_n))$$
$$= \lim_n (f(x_n) + g(y_n)) = f(z_0) + g(z_0),$$

which is not possible. Hence $f(z_0) = \liminf_n f(x_n)$ and $g(z_0) = \liminf_n g(y_n)$. Passing to a subsequence again, we may choose both sequences such that $f(z_0) = \lim_n f(x_n)$ and $g(z_0) = \lim_n g(y_n)$.   □

Then, do we have an Exact Sum Rule for the limiting subdifferential? The answer is no!

Example.   The real functions $f(t) = |t|$ and $g(t) = -|t|$ prove that the other inclusion in no longer true for the limiting subdifferential, since

$$\partial_L f(0) + \partial_L g(0) = [-1, 1] + \{-1, 1\} \not\subset \{0\} = \partial_L (f + g)(0).$$

Let us observe that both functions are Lipschitz, in other words very nice functions.

In order to have an Exact Sum Rule, the strategy is to look for conditions that guarantee that the Frechet and the limiting subdifferentials agree, which happens if the function is convex for instance. But we will deal with this question later on.

5.2 CONSTRAINED MINIMA

The Sum Rule allows us to study constrained minima of functions. Let us consider a lsc function $f : \mathbf{R}^n \to (-\infty, +\infty]$, and a closed subset S of $\mathbf{R}^n$ such that $dom f \cap S \neq \emptyset$. If f has a minimum, when restricted to S at x_0, then this point is a minimum of the auxiliary function $f + \delta_S$, and consequently $0 \in \partial(f + \delta_S)(x_0)$. If the following inclusion:

$$\partial(f + \delta_S)(x_0) \subset \partial f(x_0) + \partial \delta_S(x_0) \tag{5.1}$$

holds, we can deduce that $-\partial \delta_S(x_0) \cap \partial f(x_0) \neq \emptyset$. Under which hypotheses do we have 5.1? We present now some Lagrange Multipliers results.

PROPOSITION 5.5. *If a lsc function $f : \mathbf{R}^n \to (-\infty, +\infty]$ attains a minimum when restricted to a closed set S at a point x_0 and it is differentiable at that point, then $-\nabla f(x_0) \in \hat{N}_S(x_0)$.*

PROOF. The result is a direct consequence of Proposition 4.19 which gives us an Exact Sum Rule, and formula 4.10 which characterizes $\partial \delta_S(x_0)$ for boundary points; because if $x_0 \in int\, S$, then the result is trivial since $\nabla f(x_0) = 0$ which always belong to $\hat{N}_S(x_0)$. $\square$

Let us observe that this proposition encloses both cases $x_0 \in int\, S$ and $x_0 \in \partial S$. If the set S is a regular manifold, it is defined near x_0 as the set of zeros of a C^1 function $F = (F_1, \ldots, F_k)$, with $rang\, DF(x_0) = k$; in this case $\hat{N}\, S(x_0)$ is the linear space generated by the vectors $\nabla F_1(x_0), \ldots, \nabla F_k(x_0)$, and we have the classical Lagrange Multipliers Rule.

COROLLARY 5.6. *Let $M \subset \mathbf{R}^n$ be a smooth manifold defined in a neighborhood of x_0, V, as $M \cap V = \{x \in V : F(x) = 0\}$, with $F : V \to \mathbf{R}^k$ a C^1 function. If $f : V \to \mathbf{R}$ is a differentiable function that attains a minimum at x_0, then*

$$\nabla f(x_0) = \lambda_1 \nabla F_1(x_0) + \cdots \lambda_k \nabla F_k(x_0)$$

for real numbers $\lambda_1 \cdots \lambda_k$.

In absence of differentiability, we do not have an Exact Sum Rule in general, but we can invoke Proposition 5.4. Consequently we should characterize $\partial_L \delta_S(x_0)$.

Example. If $\zeta \in \partial_L \delta_S(x_0)$, then there exists a sequence $\{x_n\}$ converging to x_0, contained necessarily in S since $0 = \delta_S(x_0) = \lim_n \delta_S(x_n)$, and a sequence $\{\zeta_n\}$ converging to ζ, with $\zeta_n \in \partial \delta_S(x_n) = \hat{N}_S(x_n)$. We denote the set of all such vectors ζ by $N_S(x_0)$, and therefore we have $\partial_L \delta_S(x_0) = N_S(x_0)$.

DEFINITION 5.7. For a closed set $S \subset \mathbf{R}^n$, and a point $x_0 \in S$, we define the general normal cone to S at x_0 as

$$N_S(x_0) = \{v = \lim_n v_n : v_n \in \hat{N}_S(x_n), \quad \text{with} \quad x_n \in S \text{ for every } n$$

and $\lim_n x_n = x_0\}$. (See Fig. 5.1)

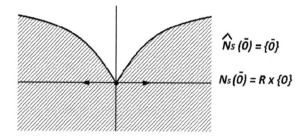

$$\widehat{N}_S(\bar{0}) = \{\bar{0}\}$$

$$N_S(\bar{0}) = R \times \{0\}$$

Fig. 5.1 Normal cone and regular normal cone may be different.

The following proposition is an immediate consequence of Proposition 5.4 and the above definition.

PROPOSITION 5.8. *Let $f : \mathbf{R}^n \to (-\infty, +\infty]$ be a lsc function, which is Lipschitz in a neighborhood of a point $x_0 \in int(dom f)$. Let $S \subset \mathbf{R}^n$ be a closed set such that $x_0 \in S$. If f attains a minimum at x_0 when restricted to S, then*

$$0 \in N_S(x_0) + \partial_L f(x_0).$$

5.3 CHAIN RULE

Now it is turn for the Chain Rule. We have not provided a subdifferential for vector functions, and although it is possible to avoid this problem and obtain a very general Chain Rule, we will only work in the restrictive setting of a first differentiable function.

PROPOSITION 5.9. *Let $f : \mathbf{R}^m \to (-\infty, +\infty]$ and $F : \mathbf{R}^n \to \mathbf{R}^m$ be two functions, lsc and differentiable respectively. Let $x_0 \in \mathbf{R}^n$ be such that $F(x_0) \in dom f$ and $\zeta \in \partial f(F(x_0))$. Then we have that*

$$DF(x_0)^*(\zeta) \in \partial(f \circ F)(x_0).$$

PROOF. Let $\zeta \in \partial f(F(x_0))$, there exists a differentiable function $\varphi : \mathbf{R}^m \to \mathbf{R}$ such that $\varphi(F(x_0)) = f(F(x_0))$, $\nabla\varphi(F(x_0)) = \zeta$ and $\varphi \leq f$ in a neighborhood of $F(x_0)$. The function $\varphi \circ F$ is differentiable, $(\varphi \circ F)(x_0) = (f \circ F)(x_0)$, and $\varphi \circ F \leq f \circ F$ in a neighborhood of x_0. Hence $DF(x_0)^*(\zeta) = \nabla(\varphi \circ F)(x_0) \in \partial(f \circ F)(x_0)$.   □

The formula above can be written also as $DF(x_0)^*(\partial f(F(x_0))) \subset \partial(f \circ F)(x_0)$. The natural question that now arises is: is it possible to represent the subgradients of the composition $f \circ F$ at x_0 as from $DF(x_0)$ and an element of $\partial f(F(x_0))$? The following example proves that in general the answer is no.

Example. Let $F(x) = 0$ for every $x \in \mathbf{R}$ and $f(t) = -|t|$ for every $t \in \mathbf{R}$. $0 \in \partial(f \circ F)(0)$, but $\partial f(F(0)) = \emptyset$.

It is not difficult to see that an exact Chain Rule is true under some restrictions on the function f, convexity for instance. We will prove a more general result later, but before that we are going to prove the opposite inclusion for the limiting subdifferential. We start, as we did for the Sum Rule, with a Fuzzy rule.

THEOREM 5.10. *Let $f : \mathbf{R}^m \to (-\infty, +\infty]$ be a lsc function and $F : \mathbf{R}^n \to \mathbf{R}^m$ a C^1 function. Let $x_0 \in \mathbf{R}^n$ be such that $F(x_0) \in \mathrm{dom} f$. Assume that f is locally Lipschitz near $F(x_0)$. Then we have that for every $\zeta \in \partial(f \circ F)(x_0)$ and every $\varepsilon > 0$, there exist $y_0 \in B(F(x_0), \varepsilon)$ and $\xi \in \partial f(y_0)$ such that*

$$|DF(x_0)^*(\xi) - \zeta| < \varepsilon.$$

PROOF. Given $\zeta \in \partial(f \circ F)(x_0)$, there is a C^1 function φ, such that $\zeta = \nabla\varphi(x_0)$ and $f \circ F - \varphi$ attains a local minimum at x_0. Let $S = \mathrm{graph} F \subset \mathbf{R}^{n+m}$, we define the auxiliary function $\phi : \mathbf{R}^n \times \mathbf{R}^m \to (-\infty, +\infty]$ by

$$\phi(x, y) = f(y) - \varphi(x) + \delta_S(x, y).$$

ϕ attains a local minimum at $(x_0, F(x_0))$, hence $(0, 0) \in \partial\phi(x_0, F(x_0))$. We deduce from Theorem 5.2 that for every $r > 0$ there exist (x_1, y_1), $(x_2, y_2) \in B((x_0, F(x_0)), r)$ such that $|\delta_S(x_0, F(x_0)) - \delta_S(x_2, y_2)| < r$ and

$$(0, 0) \in \partial(f - \varphi)(x_1, y_1) + \partial\delta_S(x_2, y_2) + rB$$
$$= \{-\nabla\varphi(x_1)\} \times \partial f(y_1) + \partial\delta_S(x_2, y_2) + rB.$$

$|\delta_S(x_0, F(x_0)) - \delta_S(x_2, y_2)| < r$ is equivalent to $y_2 = F(x_2)$. On the other hand from the characterization of the subdifferential of the indicator function, we have $\partial\delta_S(x_2, F(x_2)) = \hat{N}_S(x_2, F(x_2))$, hence the formula reads

$$(0, 0) \in \{-\nabla\varphi(x_1)\} \times \partial f(y_1) + \hat{N}_S(x_2, F(x_2)) + rB.$$

From elementary calculus, we know that $T_S(x_2, F(x_2)) = \{(v, DF(x_2)$ $(v)) : v \in \mathbf{R}^n\}$, by Proposition 4.15 we have that $(\zeta_1, \zeta_2) \in \hat{N}_S(x_2, F(x_2))$ if and only if $\langle\zeta_1, v\rangle \le -\langle\zeta_2, DF(x_2)(v)\rangle = \langle -DF(x_2)^*(\zeta_2), v\rangle$ for every $v \in \mathbf{R}^n$, equivalently if and only if $\zeta_1 = -DF(x_2)^*(\zeta_2)$. We split the formula in two, and deduce that there exists a vector $\zeta_2 \in \mathbf{R}^m$ such that

$$0 \in -\nabla\varphi(x_1) - DF(x_2)^*(\zeta_2) + rB \quad \text{and} \quad 0 \in \partial f(y_1) + \zeta_2 + rB$$

in other words, if we denote $z_0 = x_1$, $y_0 = y_1$ and change ζ_2 for $-\zeta_2$, we have $|DF(x_2)^*(\zeta_2) - \nabla\varphi(z_0)| < r$ and $\zeta_2 \in \partial f(y_0) + rB$.

Let $\xi \in \partial f(y_0)$ be such that $|\xi - \zeta_2| < r$, assume that r is chosen small enough to guarantee that $|DF(x_0)^*(\xi) - DF(x_2)^*(\zeta_2)| < \frac{\varepsilon}{3}$. This is possible since

$$|DF(x_0)^*(\xi) - DF(x_2)^*(\zeta_2)|$$
$$\leq |DF(x_0)^*(\xi) - DF(x_2)^*(\xi)| + |DF(x_2)^*(\xi) - DF(x_2)^*(\zeta_2)|$$
$$\leq |DF(x_0)^* - DF(x_2)^*||\xi| + |DF(x_2)^*||\xi - \zeta_2|.$$

The fact that f is Lipschitz near $F(x_0)$ allows us to assume that $\partial f(y_0)$ is bounded, while F being C^1 guarantees that $|DF(x_0)^* - DF(x_2)^*|$ is small, and that $|DF(x_2)^*|$ remains bounded.

On the other hand, taking r small again, we have the inequality $|\nabla\varphi(z_0) - \zeta| = |\nabla\varphi(z_0) - \nabla\varphi(x_0)| < \frac{\varepsilon}{3}$, since φ is C^1. If we choose $r < \frac{\varepsilon}{3}$ from the beginning, we get

$$|DF(x_0)^*(\xi) - \zeta| \leq |DF(x_0)^*(\xi) - DF(x_2)^*(\zeta_2)|$$
$$+|DF(x_2)^*(\zeta_2) - \nabla\varphi(z_0)| + |\nabla\varphi(z_0) - \zeta| \leq \frac{\varepsilon}{3} + r + \frac{\varepsilon}{3} < \varepsilon. \quad \square$$

In the same way as in the Sum Rule case, the limiting subdifferential provides us with an exact inclusion.

PROPOSITION 5.11. *Let $f : \mathbf{R}^m \to (-\infty, +\infty]$ be a lsc function and $F : \mathbf{R}^n \to \mathbf{R}^m$ a C^1 function. Let $x_0 \in \mathbf{R}^n$ be such that $F(x_0) \in \text{dom} f$. Assume that f is locally Lipschitz near $F(x_0)$. Then we have that*

$$\partial_L(f \circ F)(x_0) \subset DF(x_0)^*(\partial_L f(F(x_0))).$$

PROOF. Let $\zeta \in \partial_L(f \circ F)(x_0)$, there exists a sequence $\{x_n\}$ converging to x_0 and a sequence $\{\zeta_n\} \in \partial(f \circ F)(x_n)$ converging to ζ. For every n we apply Theorem 5.10 and we obtain y_n such that $|F(x_n) - y_n| < \frac{1}{n}$, and $\xi_n \in \partial f(y_n)$ such that $|DF(x_n)^*(\xi_n) - \zeta_n| < \frac{1}{n}$. It is immediate to see that

$$\lim_n y_n = F(x_0), \quad \lim_n f(y_n) = (f \circ F)(x_0),$$

and we may assume, passing to a subsequence if necessary, that $\{\xi_n\}$, which is bounded since f is Lipschitz near $F(x_0)$, converges to a vector ξ. From all these limits we deduce that $\xi \in \partial_L f(F(x_0))$. In order to finish it only remains to observe that

$$DF(x_0)^*(\xi) = \lim_n DF(x_n)^*(\xi_n) = \zeta. \qquad \square$$

Now we may wonder: is the inclusion in the proposition above an equality? Again, in general the answer is no. Let us illustrate it with an example.

Example. Let $F : \mathbf{R} \to \mathbf{R}^2$ and $f : \mathbf{R}^2 \to \mathbf{R}$ be defined by $F(t) = (t, t)$ and $f(x, y) = -|(x, y)|$. We have $DF(0) = (1, 1)$ and $\partial_L f(0, 0) = \{(\cos r, \sin r) : r \in [0, 2\pi]\}$. Hence

$$DF(0)^*(\partial_L f(0, 0)) = \{\cos r + \sin r : r \in [0, 2\pi]\} = [-\sqrt{2}, \sqrt{2}].$$

On the other hand $(f \circ F)(t) = -\sqrt{2}|t|$, and therefore $\partial_L(f \circ F)(0) = \{-\sqrt{2}, \sqrt{2}\}$.

Resuming the Chain Rule type results, we observe that as well as for the Sum Rule, one inclusion works for the Frechet subdifferential, while the other one is true for the limiting subdifferential. Can we expect, assuming more conditions on the functions, an exact equality? We first study conditions on the function F, assume for instance that $n = m$ and that F has nonzero Jacobian at x_0, we may write $f = (f \circ F) \circ F^{-1}$ near $F(x_0)$ since F is locally invertible. A direct application of Proposition 5.9 allows us to deduce

$$\begin{aligned}
&[DF(x_0)^{-1}]^*(\partial(f \circ F)(x_0)) \\
&= DF^{-1}(F(x_0))^*(\partial(f \circ F)(x_0)) \subset \partial f(F(x_0)).
\end{aligned}$$

Joining this remark with Proposition 5.9, we obtain

PROPOSITION 5.12. *Let $f : \mathbf{R}^n \to (-\infty, +\infty]$ be a lsc function, $F : \mathbf{R}^n \to \mathbf{R}^n$ a C^1 function with nonzero Jacobian at x_0. Let $x_0 \in \mathbf{R}^n$ be such that $F(x_0) \in \mathrm{dom} f$. Then we have that*

$$\partial(f \circ F)(x_0) = [DF(x_0)]^*(\partial f(F(x_0))).$$

Similarly, we have

PROPOSITION 5.13. *Let* $f : \mathbf{R}^n \to (-\infty, +\infty]$ *be a lsc function,* $F : \mathbf{R}^n \to \mathbf{R}^n$ *a* C^1 *function. Let* $x_0 \in \mathbf{R}^n$ *be such that* $F(x_0) \in domf$. *Assume that* f *is locally Lipschitz near* $F(x_0)$, *and that* F *has nonzero Jacobian at* x_0. *Then we have that*

$$\partial_L (f \circ F)(x_0) = DF(x_0)^*(\partial_L f(F(x_0))).$$

5.4 REGULAR FUNCTIONS: ELEMENTARY PROPERTIES

In order to obtain exact results concerning the sum and Chain Rules, we can also look for conditions on the second acting function, that is the function $f : \mathbf{R}^n \to (-\infty, +\infty]$. We start with a definition

DEFINITION 5.14. Let $f : \mathbf{R}^n \to (-\infty, +\infty]$ be a *lsc* function, we say that f is regular at $x_0 \in domf$ if $\partial f(x_0) = \partial_L f(x_0)$.

It is clear that C^1 functions are regular, but amazingly differential functions may lack this property.

Example. Let $f(t) = t^2 \sin \frac{1}{t}$ for $t \in \mathbf{R}, t \neq 0$, and $f(0) = 0$. f is differentiable and $f'(t) = 2t \sin \frac{1}{t} - \cos \frac{1}{t}$ for $t \neq 0$ but $f'(0) = 0$, hence f is not C^1. For this function we have

$$\partial f(0) = \{f'(0)\} = \{0\} \subset [-1, 1] = \partial_L f(0).$$

Regular functions satisfy equalities for the Sum Rule as well as for the Chain Rule. In order to prove these exact rules it is enough to combine the inclusions obtained for the Frechet and the limiting subdifferentials. Let us state the corresponding results.

PROPOSITION 5.15. *Let* $f, g : \mathbf{R}^n \to (-\infty, +\infty]$ *be two lsc functions,* $z_0 \in domf \cap domg$, *assume that one of the functions is Lipschitz in a neighborhood of* z_0, *and that both functions are regular at* z_0. *Then*

$$\partial(f + g)(z_0) = \partial f(z_0) + \partial g(z_0).$$

PROPOSITION 5.16. *Let* $f : \mathbf{R}^m \to (-\infty, +\infty]$ *be a lsc function,* $F : \mathbf{R}^n \to \mathbf{R}^m$ *a* C^1 *function. Let* $x_0 \in \mathbf{R}^n$ *be such that* $F(x_0) \in domf$. *Assume that* f *is locally Lipschitz near* $F(x_0)$ *and regular at* $F(x_0)$. *Then we have that*

$$\partial(f \circ F)(x_0) = DF(x_0)^*(\partial f(F(x_0))).$$

How large is the class of regular functions? We will develop a deeper study of this problem in the next chapter, however it is easy to see that convex functions are regular.

PROPOSITION 5.17.   *Every lsc convex function $f : \mathbf{R}^m \to (-\infty, +\infty]$ is regular at every point $x_0 \in domf$.*

PROOF.   Let $\zeta \in \partial_L f(x_0)$, there exist sequences $\{x_n\} \subset domf$ converging to x_0 and $\{\zeta_n\}$ converging to ζ such that $\zeta_n \in \partial f(x_n)$. This implies

$$f(x_n) + \langle \zeta_n, x - x_n \rangle \le f(x) \quad \text{for every} \quad x \in \mathbf{R}^n.$$

Making n tend to ∞ for every $x \in \mathbf{R}^n$ we get $f(x_0) + \langle \zeta, x - x_0 \rangle \le f(x)$. Hence $\zeta \in \partial f(x_0)$.   □

The next proposition summarizes some stability properties of regular functions. However, before that, let us observe that for a regular function f, $-f$ may lack regularity; the function $f(t) = |t|$ provides a good example.

PROPOSITION 5.18.   *The following properties hold:*

(i)   *Let $f : \mathbf{R}^n \to (-\infty, +\infty]$ be a lsc function, $x_0 \in domf$, $\lambda > 0$, if f is regular at x_0 then λf is regular at x_0 too.*

(ii)   *Let $f, g : \mathbf{R}^n \to (-\infty, +\infty]$ be two lsc functions, $x_0 \in domf \cap domg$, assume that one of the functions is Lipschitz in a neighborhood of x_0, and that both functions are regular at x_0. Then the function $f+g$ is regular at x_0.*

(iii)   *Let $f : \mathbf{R}^m \to (-\infty, +\infty]$ be a lsc function, $F : \mathbf{R}^n \to \mathbf{R}^m$ a C^1 function. Let $x_0 \in \mathbf{R}^n$ be such that $F(x_0) \in domf$. Assume that f is locally Lipschitz near $F(x_0)$ and regular at $F(x_0)$. Then we have that $f \circ F$ is regular at x_0.*

(iv)   *Let $F : \mathbf{R}^n \to \mathbf{R}^m$ be a C^1 function. Then we have that $|F|$ is regular at every point $x_0 \in \mathbf{R}^n$.*

PROOF.   (i) is trivial from the definition since $\partial(\lambda f)(x_0) = \lambda \partial f(x_0)$ and $\partial_L(\lambda f)(x_0) = \lambda \partial_L f(x_0)$. For (ii) we have to observe that

$$\partial_L(f+g)(x_0) \subset \partial_L f(x_0) + \partial_L g(x_0) = \partial f(x_0) + \partial g(x_0) \subset \partial(f+g)(x_0),$$

the first inclusion being a consequence of Proposition 5.4, the second one of Proposition 4.20, while the equality follows from regularity of f and g.

The proof of (iii) is similar, invoking Proposition 5.9 and Proposition 5.11. Finally, to get (iv) we apply (iii) with $f(y) = |y|$, which is convex and consequently locally Lipschitz and regular.   □

Let us observe that in general the composition of two regular functions is not regular, even if both functions are convex. The real convex functions $f(t) = |t|$ and $g(t) = |t| - 1$ provide us with an easy example of how $f \circ g$ may fail to be regular at 0.

In the next example we are going to look at the behavior of the maximum of C^1 functions.

Example. Let $f_1, \ldots, f_m : \mathbf{R}^m \to \mathbf{R}$ be C^1 functions. The aim of this example is to study the subdifferential of the function $g : \mathbf{R}^n \to \mathbf{R}$ defined by

$$g(x) = \max_{i=1,\ldots,m} f_i(x).$$

First of all we observe that the function $f : \mathbf{R}^m \to \mathbf{R}$ defined by $f(y) = \max\{y_1, \ldots, y_m\}$ is Lipschitz and regular since it is convex, hence g is regular, the Frechet and the limiting subdifferentials agree and we can apply Proposition 5.16. Hence $\partial g(x_0) = \partial(f \circ F)(x_0) = DF(x_0)^*(\partial f(F(x_0)))$, where $F = (f_1, \ldots, f_m)$.

In Chapter 3, we characterize the subdifferential of f as

$$\partial f(y) = \left\{ \zeta : \sum_{i \in I} \zeta_i = 1, \quad \zeta_i \geq 0, \quad \text{and } \zeta_i = 0 \text{ for } i \notin I \right\},$$

where I is the set of indices i for which $f(y) = y_i$. Consequently the subgradients of g have the following form:

$$\left(\sum_{i \in I} r_i \frac{\partial f_i}{\partial x_1}, \ldots, \sum_{i \in I} r_i \frac{\partial f_i}{\partial x_n} \right)$$

with $\sum_{i=1,\ldots,m} r_i = 1, r_i \geq 0$ for every $i = 1, \ldots, m$ and $r_i = 0$ if $i \notin I$ where $I = \{i : f_i(x_0) = g(x_0)\}$. In other words:

$$\partial g(x) = co\{\nabla f_i(x) : i \in I\}.$$

5.5 MEAN VALUE RESULTS

Our next goal is to obtain Mean Value Theorem type results. A first consideration: we cannot expect an optimal result. It is enough to think of our old friend: the function $f(t) = -|t|$ in the interval $[-1, 1]$. A Mean Value Theorem would lead us to

$$0 = f(1) - f(-1) = r(1 - (-1)) = r \cdot 2$$

which implies $r = 0$, but 0 is not a subgradient, no matter the kind, of f at any point. However we do have some interesting partial results.

PROPOSITION 5.19. *Let* $f : \mathbf{R}^n \to (-\infty, +\infty]$ *be a lsc function, locally Lipschitz around a line segment* $[x, y]$. *For every* $\varepsilon > 0$ *there are a point* $z \in [x, y] + \varepsilon B$ *and a vector* $\zeta \in \partial f(z)$ *such that*

$$f(y) - f(x) \le \langle \zeta, y - x \rangle + \varepsilon.$$

PROOF. Let us consider the function $F : \mathbf{R} \to \mathbf{R}^n$ defined by $F(t) = ty + (1 - t)x$. The function $h(t) = f(F(t)) - tf(y) - (1 - t)f(x)$ is Lipschitz in $[0, 1]$, hence it attains a minimum at a point $t_0 \in [0, 1]$. If $t_0 \in (0, 1)$ we have

$$0 \in \partial h(t_0) = f(x) - f(y) + \partial(f \circ F)(t_0)$$

and consequently $f(y) - f(x) \in \partial(f \circ F)(t_0)$. The function F is C^1, while f is locally Lipschitz near $F(t_0)$. We can apply Theorem 5.10 and deduce that there are a point $z \in B(F(t_0), \varepsilon)$ and a $\zeta \in \partial f(z)$ such that

$$f(y) - f(x) - \langle \zeta, y - x \rangle \le |\langle \zeta, y - x \rangle - (f(y) - f(x))|$$
$$= |DF(t_0)^*(\zeta) - (f(y) - f(x))| < \varepsilon.$$

If $t_0 = 0$ or $t_0 = 1$ then $h(t) \ge h(0) = h(1) = 0$ for every $t \in [0, 1]$. Let $\delta > 0$ be such that $h(t) < \frac{\varepsilon}{4}$ if $t \in [0, \delta]$, and let us consider a differentiable function $\varphi : [0, 1] \to \mathbf{R}$ satisfying the following conditions: $\varphi(0) = \varphi(1) = 0$, $\varphi'(t) < \frac{\varepsilon}{2}$ for every $t \in [0, 1]$, and φ attains a minimum at δ with $\varphi(\delta) = -\frac{\varepsilon}{4}$ (this is possible if we choose $\delta < \frac{\varepsilon}{4}$ for instance, see Fig. 5.2). The function $h + \varphi$ attains a minimum at a point $t_0 \in (0, 1)$ and we can repeat the argument above, hence

$$0 \in \partial(h + \varphi)(t_0) = f(x) - f(y) + \varphi'(t_0)$$
$$+ \partial(f \circ F)(t_0) < f(x) - f(y) + \frac{\varepsilon}{2} + \partial(f \circ F)(t_0)$$

invoking Theorem 5.10 again with $\frac{\varepsilon}{2}$ instead of ε we get the inequality.   □

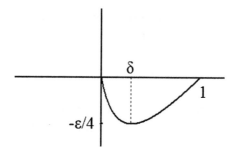

Fig. 5.2 *The auxiliary function φ.*

THEOREM 5.20 (*Mean Value Inequality*). *Let $f : \mathbf{R}^n \to (-\infty, +\infty]$ be a lsc function, locally Lipschitz around a line segment $[x, y]$, then there are a point $z_0 \in [x, y]$ and a vector $\zeta_0 \in \partial_L f(z_0)$ such that*

$$f(y) - f(x) \le \langle \zeta_0, y - x \rangle.$$

PROOF. We apply Proposition 5.19 with $\varepsilon = \frac{1}{n}$, and obtain sequences $\{z_n\}$ and $\zeta_n \in \partial f(z_n)$ such that $z_n \in [x, y] + \frac{1}{n}B$ and

$$f(y) - f(x) \le \langle \zeta_n, y - x \rangle + \frac{1}{n}.$$

The sequence $\{z_n\}$ is bounded hence, passing to a subsequence if necessary, we can assume that it converges to a point z_0 that necessarily belongs to $[x, y]$. The sequence $\{\zeta_n\}$ is bounded too since f is Lipschitz near $[x, y]$, and therefore we can assume that it also converges to a vector ζ_0 (possibly we will require another subsequence). This ζ_0 fits the conditions of limiting subgradients at z_0 (remember that f is continuous). Taking limits in the above formula we obtain the Mean Value Inequality.   □

Although it is not possible to obtain a Mean Value Equality, the proof of Proposition 5.19 suggests the following sub-super result.

PROPOSITION 5.21. *Let $f : \mathbf{R}^n \to (-\infty, +\infty]$ be a lsc function, locally Lipschitz around a line segment $[x, y]$, then there are a point $z_0 \in [x, y]$ and a vector $\zeta_0 \in \partial_L f(z_0) \cup \partial^L f(z_0)$ such that*

$$f(y) - f(x) = \langle \zeta_0, y - x \rangle.$$

PROOF. We proceed as in Proposition 5.19, defining the continuous function $h : [0, 1] \to \mathbf{R}$, $h(t) = f(ty + (1 - t)x) - tf(y) - (1 - t)f(x)$.

From continuity of h and compactness of $[0, 1]$, we deduce that h attains either a maximum or a minimum, at a point $t_0 \in (0, 1)$ since $h(0) = h(1)$. If t_0 is a minimum, we proceed as in the proof of the Proposition and there is a point $z \in [x, y] + \varepsilon B$ and $\zeta \in \partial^- f(z)$ such that $|\langle \zeta, y - x \rangle - (f(y) - f(x))| < \varepsilon$. If t_0 is a maximum, we proceed as above obtaining the same estimation but with $\zeta \in \partial^+ f(z)$. Hence we have proved that for every $\varepsilon > 0$ there exist a $z \in [x, y] + \varepsilon B$ and a $\zeta \in \partial^- f(z) \cup \partial^+ f(z)$ such that

$$|\langle \zeta, y - x \rangle - (f(y) - f(x))| < \varepsilon.$$

A convergence argument, similar to the Mean Value Inequality one, guarantees the existence of a $z_0 \in [x, y]$ and a $\zeta_0 \in \partial_L f(z_0) \cup \partial^L f(z_0)$ such that $|f(y) - f(x) - \langle \zeta_0, y - x \rangle| = 0$.   □

5.6 DECREASING FUNCTIONS

It is well known that smooth analysis is a useful tool for studying the increasing or decreasing behavior of one variable functions. Results can also be obtained in the several variables setting. In this section we will prove an easy characterization of nonincreasing one variable functions as well as the important Decrease Principle.

Let us characterize nonincreasing functions in terms of nonsmooth properties.

PROPOSITION 5.22.   *A lsc function $f : \mathbf{R} \to \mathbf{R}$ is nonincreasing if and only if $\partial f(x) \subset (-\infty, 0]$ for every $x \in \mathbf{R}$.*

PROOF.   Let us assume that f is nonincreasing. The subderivative function $df(x)(1) = \liminf_{t \downarrow 0, r \to 1} \frac{f(x+tr) - f(x)}{t}$ is clearly nonpositive, if $\zeta \in \partial f(x)$ then $\zeta.1 \le df(x)(1)$ by Lemma 4.11, hence $\zeta \le 0$.

Conversely, let $x < y$, let us fix $\varepsilon > 0$. We deduce from Proposition 5.19 that there exist a $z \in [x, y] + \varepsilon B = (x - \varepsilon, y + \varepsilon)$, and a $\zeta \in \partial f(z)$ such that

$$f(y) - f(x) \le \langle \zeta, y - x \rangle + \varepsilon \zeta(y - x) + \varepsilon \le \varepsilon.$$

By letting $\varepsilon \downarrow 0$, we get $f(y) \le f(x)$.   □

In the above Proposition, we may replace $\partial f(x)$ by $\partial_L f(x)$ since $\partial f(x) \subset (-\infty, 0]$ for every $x \in \mathbf{R}$ if and only if $\partial_L f(x) \subset (-\infty, 0]$ for

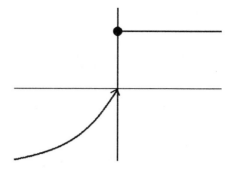

Fig. 5.3 A nondecreasing usc function

every $x \in \mathbf{R}$. For nondecreasing functions we have the following result (See Fig. 5.3).

COROLLARY 5.23. *An usc function $f : \mathbf{R} \to \mathbf{R}$ is nondecreasing if and only if $\partial^+ f(x) \subset [0, +\infty)$ for every $x \in \mathbf{R}$.*

Proposition 5.22 is a particular case of a more general result. We require some definitions.

DEFINITION 5.24. Let $C \subset \mathbf{R}^n$ be a convex compact, $f : \mathbf{R}^n \to \mathbf{R}$ a *lsc* function. We say that f is strongly decreasing with respect to C if $f(y) \leq f(x)$ for every $y \in x + tC, t \geq 0$.

For one variable functions f, it is clear that f is nonincreasing if and only if it is strongly decreasing with respect to $C = \{1\}$, while it is nondecreasing if and only if it is strongly decreasing with respect $C = \{-1\}$.

We introduce the following number in order to characterize strongly decreasing functions.

DEFINITION 5.25. For $C \subset \mathbf{R}^n$ compact and convex, and $v \in \mathbf{R}^n$, we define

$$H_C(v) = \max\{\langle v, c \rangle : c \in C\}$$

PROPOSITION 5.26. *Let $C \subset \mathbf{R}^n$ be a convex compact and $f : \mathbf{R}^n \to \mathbf{R}$ a lsc function. We have that f is strongly decreasing with respect to C if and only if $H_C(\zeta) \leq 0$ for every $x \in \mathbf{R}^n$ and $\zeta \in \partial f(x)$.*

PROOF. If f is strongly decreasing with respect to C, we have that $f(x + tc) \leq f(x)$ for every $c \in C$ and $t > 0$. Let $\zeta \in \partial f(x)$ be a subgradient of f. We have

$$0 \leq \liminf_{h \to 0} \frac{f(x+h) - f(x) - \langle \zeta, h \rangle}{|h|}$$
$$\leq \liminf_{t \downarrow 0} \frac{f(x+tc) - f(x) - t \langle \zeta, c \rangle}{t|c|},$$

hence

$$\frac{\langle \zeta, c \rangle}{|c|} \leq \liminf_{t \downarrow 0} \frac{f(x+tc) - f(x)}{t|c|} \leq 0,$$

therefore $H_C(\zeta) \leq 0$.

Conversely, given $y = x + tc$ and $\varepsilon > 0$, we have

$$f(y) - f(x) \leq \langle \zeta, y - x \rangle + \varepsilon$$

for an appropriate ζ, by Proposition 5.19. Hence

$$f(y) - f(x) \leq t \langle \zeta, c \rangle + \varepsilon \leq t H_C(\zeta) + \varepsilon \leq \varepsilon.$$

Letting $\varepsilon \to 0$ we have that $f(y) \leq f(x)$, and the result is established.   □

An important tool for studying decreasing functions are mean value results, for instance in the preceding result we invoked a Fuzzy Mean Value Inequality. But in order to achieve other results we require a stronger result. This is the *Multidirectional Mean Value Theorem*. We begin with an elementary result for differentiable functions.

PROPOSITION 5.27.   *Let $f : \mathbf{R}^n \to \mathbf{R}$ be a lsc function. Assume that f attains its minimum when restricted to a closed convex set C at $x_0 \in C$. Assume also that f is differentiable at x_0. Then*

$$\langle \nabla f(x_0), x - x_0 \rangle \geq 0$$

for every $x \in C$.

PROOF.   By Proposition 5.5, we have that $-\nabla f(x_0) \in \hat{N}_C(x_0)$ or equivalently $\langle \nabla f(x_0), v \rangle \geq 0$ for every $v \in T_C(x_0)$. The fact that for convex sets $x - x_0 \in T_S(x_0)$ provided that $x \in C$, gives us the result.   □

We are going to establish a smooth Multidirectional Mean Value Theorem, we will use the following notation in the sequel:

$$[x, C] = co(\{x\} \cup C).$$

THEOREM 5.28. *Let $C \subset \mathbf{R}^n$ be closed and convex, $x_0 \in \mathbf{R}^n$, U an open neighborhood of $[x_0, C]$, $f : U \to \mathbf{R}$ a differentiable function. Then there exists $z \in [x_0, C]$ such that*

$$\min\{f(c) : c \in C\} \le f(x_0) + \langle \nabla f(z), y - x_0 \rangle$$

for every $y \in C$.

PROOF. Let denote $m = \min\{f(c) : c \in C\} - f(x_0)$. We define the *function $g : [0, 1] \times U \to \mathbf{R}$ by*

$$g(t, y) = f(x_0 + t(y - x_0)) - mt$$

g is continuous, hence it attains a minimum over the compact $[0, 1] \times C$ at a point (t_0, c_0). We consider three different cases.

If $t_0 \in (0, 1)$, we define $z = x_0 + t_0(c_0 - x_0) \in [x_0, C]$. The function $h_1(t) = g(t, c_0)$ attains a minimum at t_0, hence

$$0 = h_1'(t_0) = \langle \nabla f(x_0 + t_0(c_0 - x_0)), c_0 - x_0 \rangle - m = \langle \nabla f(z), c_0 - x_0 \rangle - m. \tag{5.2}$$

On the other hand, $h_2(y) = g(t_0, y)$ attains a minimum over C at c_0, hence, by Proposition 5.27, we have that

$$0 \le \langle \nabla h_2(c_0), y - c_0 \rangle = \langle t_0 \nabla f(z), y - c_0 \rangle \tag{5.3}$$

for every $y \in C$. Joining 5.2 and 5.3 we obtain

$$m \le \langle \nabla f(z), y - c_0 \rangle + \langle \nabla f(z), c_0 - x_0 \rangle = \langle \nabla f(z), y - x_0 \rangle$$

for every $y \in C$, which establishes the result.

If $t_0 = 0$, we define $z = x_0$, and observe that $f(x_0) = g(0, c_0) \leq g(t, y) = f(x_0 + t(y - x_0)) - mt$, hence

$$m \leq \frac{f(x_0 + t(y - x_0)) - f(x_0)}{t}$$

and letting $t \downarrow 0$ we obtain $m \leq \langle \nabla f(x_0), y - x_0 \rangle$ for every $y \in C$.

Finally, if $t = 1$, we define $z = x_0$ again, the following inequalities

$$f(x_0) = g(0, c_0) \geq g(1, c_0) = f(c_0) - m$$
$$= f(c_0) - \min\{f(c) : c \in C\} + f(x_0) \geq f(x_0)$$

imply that $g(0, c_0) = g(1, c_0)$, and we are in the case $t_0 = 0$ again.   □

The Multidirectional Mean Value Theorem has many interesting consequences, the following corollary is one of them.

COROLLARY 5.29.   *Let $f : U \to \mathbf{R}$ be a differentiable function, where U is an open neighborhood of the closed unit ball $\overline{B}$. If $|\nabla f(x)| \geq 1$ for every $x \in \overline{B}$, then*

$$\max\{f(x) : x \in \overline{B}\} - \min\{f(x) : x \in \overline{B}\} \geq 2.$$

PROOF.   We apply Theorem 5.28 with $x_0 = 0$ and $C = \overline{B}$ and obtain $\min\{f(x) : x \in \overline{B}\} - f(0) \leq \langle \nabla f(z), y \rangle$ for every $y \in \overline{B}$, hence in particular

$$\min\{f(x) : x \in \overline{B}\} - f(0) \leq -|\nabla f(z)| \leq -1$$

Replacing f by $-f$, we have

$$-\max\{f(x) : x \in \overline{B}\} + f(0) \leq -1.$$

Joining both inequalities we achieve the result.   □

Let us observe that the result is no longer true if differentiability fails, even at only one point; consider the function $f(x) = |x|$ for instance.

We will present the nonsmooth version of the Multidirectional Mean Value Theorem without proof.

THEOREM 5.30. *Let $C \subset \mathbf{R}^n$ be compact and convex, $x_0 \in \mathbf{R}^n$, $\varepsilon > 0$, and $f : \mathbf{R}^n \to \mathbf{R}$ a lsc function. Then there exist $z \in [x_0, C] + \varepsilon B$ and $\zeta \in \partial f(z)$ such that*

$$\min\{f(c) : c \in C\} \le f(x_0) + \langle \zeta, y - x_0 \rangle$$

for every $y \in C$.

The following example proves that we cannot guarantee that $z \in [x_0, C]$.

Example. The function $f : \mathbf{R} \to \mathbf{R}$ defined by $f(x) = -\sqrt{-x}$ if $x \le 0$ and $f(x) = 1$ otherwise, together with the point $x_0 = 0$ and the set $C = \{1\}$, show that the restriction $z \in [x_0, C] + \varepsilon B = (-\varepsilon, 1 + \varepsilon)$, instead of $z \in [x_0, C] = [0, 1]$, cannot be avoided.

Theorem 5.30 allows us to characterize another class of decreasing functions. Let us introduce some definitions.

DEFINITION 5.31. Let $C \subset \mathbf{R}^n$ be a convex compact set, $f : \mathbf{R}^n \to \mathbf{R}$ a *lsc* function. We say that f is weakly decreasing with respect to C if for every $t \ge 0$, there exists $y \in x + tC$, such that $f(y) \le f(x)$.

For a one variable function f, it is clear that f is nonincreasing if and only if it is weakly decreasing with respect to $C = \{1\}$, while it is nondecreasing if and only if it is weakly decreasing with respect $C = \{-1\}$.

As we did for strongly decreasing functions, we introduce a number in order to characterize weakly decreasing functions.

DEFINITION 5.32. For $C \subset \mathbf{R}^n$ compact and convex, and $v \in \mathbf{R}^n$, we define

$$h_C(v) = \min\{\langle v, c \rangle : c \in C\}.$$

PROPOSITION 5.33. *Let $C \subset \mathbf{R}^n$ be a convex compact and $f : \mathbf{R}^n \to \mathbf{R}$ a lsc function. We have that f is weakly decreasing with respect to C if and only if $h_C(\zeta) \le 0$ for every $x \in \mathbf{R}^n$ and $\zeta \in \partial f(x)$.*

PROOF. Let us suppose first that f is weakly decreasing, let $\zeta \in \partial f(x)$. For every $t > 0$, there exists a $c \in C$ such that $f(x + tc) \le f(x)$. Let us consider a sequence $t_n \downarrow 0$, we consider the corresponding c_n, then we have

$$\liminf_n \frac{f(x + t_n c_n) - f(x) - \langle \zeta, t_n c_n \rangle}{t_n |c_n|} \geq \liminf_{h \to 0} \frac{f(x + h) - f(x) - \langle \zeta, h \rangle}{|h|} \geq 0.$$

We may assume without loss of generality that the sequence $\{c_n\}$ converges to $c \in C$ by compactness, hence

$$\liminf_n \frac{f(x + t_n c_n) - f(x)}{t_n |c_n|} - \frac{\langle \zeta, c \rangle}{|c|} = \liminf_n \left(\frac{f(x + t_n c_n) - f(x)}{t_n |c_n|} - \frac{\langle \zeta, c_n \rangle}{|c_n|} \right) \geq 0.$$

From the choice of c_n we deduce

$$\frac{\langle \zeta, c \rangle}{|c|} \leq \liminf_n \frac{f(x + t_n c_n) - f(x)}{t_n |c_n|} \leq 0$$

and consequently $h_C(\zeta) \leq 0$.

Conversely, fix x, let $t > 0$, we apply Theorem 5.30 to the point x and the compact convex $A = x + tC$, we have that there are a z and a $\zeta \in \partial f(z)$ such that

$$\min\{f(x + tc) : c \in C\} - f(x) \leq \langle \zeta, tc \rangle$$

for every $c \in C$, hence

$$\min\{f(x + tc) : c \in C\} - f(x) \leq t h_C(\zeta) \leq 0.$$

We deduce that there exists a $c \in C$ such that $f(x + tc) \leq f(x)$, and consequently f is weakly decreasing. □

We will now establish the *Decrease Principle*.

THEOREM 5.34. *Let $f : \mathbf{R}^n \to \mathbf{R}$ be a lsc function. Assume that there is a positive constant a such that $|\zeta| \geq a$ for every $\zeta \in \partial f(x)$, $x \in \overline{B}(x_0, r_0)$. Then*

$$\min\{f(x) : x \in \overline{B}(x_0, r_0)\} \leq f(x_0) - ar_0.$$

PROOF. We apply Theorem 5.30 with $C = \overline{B}(x_0, r_0 - \varepsilon) = [x_0, C]$. We deduce

$$\min\{f(x) : x \in \overline{B}(x_0, r_0 - \varepsilon)\} - f(x_0) \leq \langle \zeta, y - x_0 \rangle$$

for every $y \in \overline{B}(x_0, r_0 - \varepsilon)$, where ζ is a subgradient at a point $z \in \overline{B}(x_0, r_0)$. Particularizing for $y = x_0 - (r_0 - \varepsilon)\frac{\zeta}{|\zeta|}$ we obtain

$$\min\{f(x) : x \in \overline{B}(x_0, r_0 - \varepsilon)\} - f(x_0) \leq -(r_0 - \varepsilon)|\zeta| \leq -a(r_0 - \varepsilon),$$

and letting $\varepsilon \downarrow 0$ we get the result.   □

5.7 PROBLEMS

(1) Prove that $\partial_L f(x_0)$ is closed.

(2) Give an example of a function f and a point x_0 such that $\partial_L f(x_0) = \emptyset$.

(3) Prove that $\partial_L f(x_0) \neq \emptyset$ provided that $f : \mathbf{R}^n \to (-\infty, +\infty]$ is Lipschitz near $x_0 \in dom f$.

(4) Give an example of two continuous functions $f, g : \mathbf{R} \to \mathbf{R}$, a $\zeta \in \partial(f+g)(0)$, and a positive ε such that $\zeta \notin \partial f(x) + \partial g(y)$ when $x, y \in (-\varepsilon, +\varepsilon)$. (Hint: Consider the function $f(t) = -|t| - t^2$).

(5) Calculate $\partial_L f(0)$ for the functions defined in problems 1 and 2 of Chapter 4.

(6) Calculate $\partial_L f(x, y)$ at every $(x, y) \in \mathbf{R}^2$, for the function $f(x, y) = |x| - |y|$.

(7) Let $f : \mathbf{R}^n \to \mathbf{R}$ be C^1 in a neighborhood of a point x_0. Prove that $\partial_L f(x_0) = \{\nabla f(x_0)\}$.

(8) Give an example of two, necessarily nonLipschitz functions, $f, g : \mathbf{R} \to \mathbf{R}$, such that $\partial_L(f + g)(0) \not\subset \partial_L f(0) + \partial_L g(0)$.

(9) Give an example of two closed sets $S_1, S_2 \subset \mathbf{R}^n$, and a point $x_0 \in S_1 \cap S_2$, such that

$$N_{S_1 \cap S_2}(x_0) \not\subset N_{S_1}(x_0) + N_{S_2}(x_0).$$

Let us observe that this formula means

$$\partial_L(\delta_{S_1} + \delta_{S_2})(x_0) \not\subset \partial_L \delta_{S_1}(x_0) + \partial_L \delta_{S_2}(x_0).$$

Is the other inclusion true?

(10) Let $S \subset \mathbf{R}^n$ be a closed set, $x_0 \in S$. Prove that $N_S(x_0)$ is a closed cone.

(11) Let $S \subset \mathbf{R}^n$ be a closed set, $x_0 \in S$. For a point $x \in \mathbf{R}^n$, we define $proj_S(x)$ as the set of all points $\bar{x} \in S$ such that $|x - \bar{x}| = d_S(x)$.

(a) Prove that $x - \bar{x} \in \hat{N}_S(\bar{x})$ for every $x \in \mathbf{R}^n$.

(b) Prove that if $x_0 \in S, \{x_n\} \subset \mathbf{R}^n$ with $\lim_n x_n = x_0$, and $\bar{x}_n \in proj_S(x_n)$ for every n, then $\lim_n \bar{x}_n = x_0$ too.

(c) Prove that the set $\{z \in \partial S : \hat{N}_S(z) \neq \{0\}\}$ is dense in ∂S.

(d) Deduce that $N_S(x_0) \neq \{0\}$ whenever $x_0 \in \partial S$.

(12) Calculate the normal cone $N_S(0, 0)$ for the sets S defined in Problem 13 of Chapter 4.

(13) Let $f : \mathbf{R}^n \to \mathbf{R}$ be a differentiable function in a neighborhood of a point x_0. Calculate the normal cone to the graph of f at $(x_0, f(x_0))$. Under which conditions do the general and regular cones agree?

(14) Let $f : \mathbf{R}^n \to (-\infty, +\infty]$ be a lsc function, $x_0 \in domf$. Prove that

$$\partial_L f(x_0) = \{\zeta \in \mathbf{R}^n : (\zeta, -1) \in N_{epif}(x_0, f(x_0))\}.$$

(15) Prove that
$$\partial_L d_S(x) = N_S(x) \cap \overline{B}(0, 1) \tag{5.4}$$

for every $x \in S$.

(16) Let $S = [0, 1]^n \subset \mathbf{R}^n$ be a n-cube. $f : \mathbf{R}^n \to \mathbf{R}$ differentiable in a neighborhood of S. Assume that $[x, x+\nabla f(x)] \cap S = \{x\}$ for every $x \in \partial S$. Prove that there is a point $x_0 \in S$ such that $\nabla f(x_0) = 0$.

(17) Let $f : \mathbf{R}^2 \to \mathbf{R}$ be differentiable in a neighborhood of the square $S = [-1, 1]^2$. Assume that

$$\frac{\partial f(x, y)}{\partial x} \frac{\partial f(x, y)}{\partial y} < 0 \text{ for every } (x, y) \in S.$$

Prove that f attains its maximum and minimum over S at vertices $(1, -1), (-1, 1)$.

(18) Let $f : \mathbf{R}^n \to \mathbf{R}$ be a Lipschitz function. For every $\alpha \in \mathbf{R}$, let us denote the α-level set $\{x \in \mathbf{R}^n : f(x) = \alpha\}$ by S_α. Prove that

$$\partial_L^+ f(x) \cap N_{S_\alpha}(x) \neq \emptyset$$

for every $x \in S_\alpha$. Give an example in order to prove that the Lipschitz condition cannot be avoided.

(19) Let $F : \mathbf{R}^n \to \mathbf{R}^n$ be a differentiable function. Assume that the function $\phi(x) = |x - F(x)|$ is differentiable too. Prove that $DF(x_0) = I$ at every fixed point x_0 of F.

(20) Let $f, g : \mathbf{R} \to \mathbf{R}$ be defined by $f(t) = -|t| + \sqrt{|t|}$ and $g(t) = -\sqrt{|t|}$. Observe that although f and g are regular, this property fails for $f + g$.

(21) Let $f : \mathbf{R}^n \to \mathbf{R}$ be defined by $f(x) = \min\{x_i : i = 1, \ldots, n\}$. Calculate $\partial f(x)$ and $\partial_L f(x)$ at every point $x \in \mathbf{R}^n$.

(22) Let $f_1, \ldots, f_m : \mathbf{R}^n \to \mathbf{R}$ be C^1 functions. Define $g : \mathbf{R}^n \to \mathbf{R}$ as $g(x) = \min\{f_i(x) : i = 1, \ldots, m\}$. Prove that

$$\partial_L g(x) = \{\nabla f_i(x) : g(x) = f_i(x)\}.$$

(23) Let S_1, S_2 be two closed disjoint sets, one of them compact. Let $x_1 \in S_1$ and $x_2 \in S_2$ be such that $d(S_1, S_2) = |x_1 - x_2|$. Prove that

$$x_2 - x_1 \in \hat{N}_{S_1}(x_1) \quad \text{and} \quad x_1 - x_2 \in \hat{N}_{S_2}(x_2).$$

(24) Let $S = \{(x, y) \in \mathbf{R}^2 : xy = 0\}$. Calculate $\partial d_S(0, 0)$ and $\partial_L d_S(0, 0)$. If we define $\varphi_1(t) = d_S(t, 0)$, is it true that $\zeta_1 \in \partial_L \varphi_1(0)$ for every $\zeta \in \partial_L d_S(0, 0)$?

(25) Let $f : \mathbf{R}^n \to \mathbf{R}$ be a C^1 function. Let $C = co\{e_1, \ldots, e_n\}$. What is the meaning of f being strongly, respectively weakly, decreasing with respect to C?

(26) Let $G = \{(x, y) \in \mathbf{R}^2 : |xy| < 1\}$, let $F : G \to \mathbf{R}^2$ be defined by $F(x, y) = (\sin xy, \cos xy)$, and $f : \mathbf{R}^2 \to \mathbf{R}$ by $f(x, y) = |x| - |y|$. Prove that f is regular at every point of $F(G)$. Calculate $\partial(f \circ F)(x_0, y_0)$ for every $(x_0, y_0) \in G$.

(27) Characterize the strongly and weakly decreasing functions with respect to the closed unit ball.

(28) Characterize the one variable functions $f : \mathbf{R} \to \mathbf{R}$ that are strongly (respectively weakly) decreasing with respect to $C = [0, 2]$. Give an example of a weakly but not strongly decreasing function with respect to that set.

(29) Prove that if a function is strongly decreasing with respect to a set C, it is also strongly decreasing with respect to every subset $S \subset C$. Prove that the result is no longer true if we replace strongly by weakly.

CHAPTER 6

Lipschitz Functions and the Generalized Gradient

Recall that a function $f : \mathbf{R}^n \to (-\infty, +\infty]$ is Lipschitz on $A \subset \mathbf{R}^n$ if there is a positive constant K such that

$$|f(x) - f(y)| \leq K|x - y| \quad \text{for every} \quad x, y \in A.$$

Let us observe that $A \subset dom f$ necessarily. Sometimes we will say that f is K-Lipschitz in order to point out that the inequality is true for the constant K. As we saw in Chapter 5, Lipschitz's property is a very useful tool to establish results in calculus. The purpose of this chapter is the in depth study of Lipschitz functions, as well as the introduction of a new nonsmooth concept that contains both subgradients and supergradients.

6.1 LIPSCHITZ REGULAR FUNCTIONS

In Chapter 5 we made an extensive use of the fact that a Lipschitz function has bounded subdifferentials. As a matter of fact this property characterizes Lipschitz functions.

THEOREM 6.1. *Let $U \subset \mathbf{R}^n$ be an open set. A lsc function $f : \mathbf{R}^n \to (-\infty, +\infty]$ is K-Lipschitz on U if and only if $|\zeta| \leq K$ for every $z \in U$ and $\zeta \in \partial f(z)$.*

PROOF. We already proved the only implication in Proposition 4.4.

Let us assume that $|\zeta| \leq K$ for every $\zeta \in \partial f(z)$. Fix $r > 0$, for every $x, y \in U, x \neq y$, we apply Proposition 5.19 with $\varepsilon \leq r|x - y|$, and such that $[x, y] + \varepsilon B \subset U$, to obtain $z \in [x, y] + \varepsilon B \subset U$ and $\zeta \in \partial f(z)$ satisfying

$$f(y) - f(x) \leq \langle \zeta, y - x \rangle + \varepsilon \leq |\zeta||y - x| + r|y - x| \leq (K + r)|y - x|.$$

As the same inequality holds changing the roles of x and y, we have $|f(x) - f(y)| \leq (K + r)|x - y|$. Letting $r \searrow 0$ we conclude that $|f(x) - f(y)| \leq K|x - y|$ if $x \neq y$ and consequently for every $x, y \in U$. $\qquad\square$

An Introduction to Nonsmooth Analysis. http://dx.doi.org/10.1016/B978-0-12-800731-0.00006-0

A consequence of this theorem is that Lipschitz functions have non empty limiting subdifferential. Let us see it.

PROPOSITION 6.2. *Let* $f : \mathbf{R}^n \to (-\infty, +\infty]$ *be locally Lipschitz around* $x_0 \in dom f$. *Then*

$$\partial_L f(x_0) \neq \emptyset.$$

PROOF. The Density Theorem allows us to choose a sequence $\{x_n\}$ converging to x_0 such that $\partial f(x_n) \neq \emptyset$, if we take $\zeta_n \in \partial f(x_n)$, we have that the sequence $\{\zeta_n\}$ is bounded since f is locally Lipschitz around x_0 and, passing to a subsequence if necessary, we may assume also that $\{\zeta_n\}$ is convergent to a vector ζ that necessarily belongs to $\partial_L f(x_0)$.   □

In Definition 4.10 we introduced the subderivative function $df(x) : \mathbf{R}^n \to [-\infty, +\infty]$ as a useful tool for the practical calculus of the subdifferential. When the function f is Lipschitz, it is possible to simplify its expression.

PROPOSITION 6.3. *Let* $f : \mathbf{R}^n \to (-\infty, +\infty]$, *and* $x_0 \in dom f$. *If* f *is locally Lipschitz around* x_0, *then*

$$df(x_0)(w_0) = \liminf_{t \searrow 0} \frac{f(x_0 + tw_0) - f(x_0)}{t}.$$

PROOF. We assume that f is K-Lipschitz in a neighborhood of x_0. In order to study the quotient $\frac{f(x_0 + tw) - f(x_0)}{t}$ we observe that

$$\frac{f(x_0 + tw) - f(x_0)}{t} = \frac{f(x_0 + tw_0) - f(x_0)}{t}$$
$$+ \frac{f(x_0 + tw) - f(x_0 + tw_0)}{t}$$

and consequently

$$-K|w - w_0| + \frac{f(x_0 + tw_0) - f(x_0)}{t} \leq \frac{f(x_0 + tw) - f(x_0)}{t}$$

and

$$\frac{f(x_0 + tw) - f(x_0)}{t} \leq K|w - w_0| + \frac{f(x_0 + tw_0) - f(x_0)}{t}.$$

Taking lower limits when $t \searrow 0$ and $w \to w_0$ we get

$$\liminf_{t \searrow 0} \frac{f(x_0 + tw_0) - f(x_0)}{t} \le df(x_0)(w_0)$$

$$\le \liminf_{t \searrow 0} \frac{f(x_0 + tw_0) - f(x_0)}{t}. \qquad \square$$

Our next goal is to study the relationship between the Lipschitz property, differentiability and regularity. As we saw above, C^1 functions are locally Lipschitz, however differentiability is not enough. Let us see an example.

Example. Let $f : \mathbf{R} \to \mathbf{R}$ be defined by $f(t) = |t|^{\frac{3}{2}} \sin \frac{1}{t}$ if $t \ne 0$ and $f(0) = 0$. It is easy to see that f is not C^1 around 0 nor regular at 0 ($\partial_L f(0) = \mathbf{R}!$) even though f is differentiable everywhere. Is it Lipschitz near 0? Let us suppose that there is a positive constant K such that $|f(x) - f(y)| \le K|x - y|$ in a neighborhood of 0. Let us take $x = (2n\pi)^{-1}$ and $y = (2n\pi + \frac{\pi}{2})^{-1}$ for an arbitrary n, we have that $|f(x) - f(y)| = (2n\pi + \frac{\pi}{2})^{-\frac{3}{2}}$ while

$$|x - y| = \frac{\frac{\pi}{2}}{\left(2n\pi + \frac{\pi}{2}\right) 2n\pi},$$

hence

$$K \ge \frac{\left(2n\pi + \frac{\pi}{2}\right) 2n\pi}{\frac{\pi}{2}} \left(2n\pi + \frac{\pi}{2}\right)^{-\frac{3}{2}} = \frac{4n}{\left(2n\pi + \frac{\pi}{2}\right)^{\frac{1}{2}}}$$

$$\ge \frac{4n}{(4n\pi)^{\frac{1}{2}}} = \frac{2\sqrt{n}}{\sqrt{\pi}} \ge \sqrt{n}$$

for every n, which is impossible. Therefore f is not locally Lipschitz near 0.

Summing up, the Lipschitz property is neither weaker nor stronger than differentiability. We introduce the next definition and attempt to characterize the Lipschitz property in terms of differentiability.

DEFINITION 6.4. A function $f : \mathbf{R}^n \to [-\infty, +\infty]$ is strictly differentiable at a point $x_0 \in \text{dom} f$ if there is a vector v such that

$$\lim_{x,y \to x_0, x \ne y} \frac{f(x) - f(y) - \langle v, x - y \rangle}{|x - y|} = 0.$$

Let us observe that strict differentiability implies differentiability, and that $v = \nabla f(x_0)$ necessarily. The next theorem will relate strict differentiability with the Lipschitz property and regularity, but we observe first that the above example also proves that differentiability at x_0 does not imply that $\partial_L f(x_0) = \{\nabla f(x_0)\}$. We require f to be C^1 for this equality, having in general $\nabla f(x_0) \in \partial_L f(x_0)$.

THEOREM 6.5. *Let $f : \mathbf{R}^n \to [-\infty, +\infty]$ be a function, $x_0 \in \operatorname{dom} f$. The following assertions are equivalent:*

(i)   *f is strictly differentiable at x_0.*
(ii) *f is locally Lipschitz around x_0 and $\partial_L f(x_0)$ is a singleton.*
(iii) *f is locally Lipschitz around x_0 and both f and $-f$ are regular at x_0.*

PROOF. (i) $\Rightarrow$ (ii). We may define the auxiliary function $g(x) = f(x) - \langle \nabla f(x_0), x \rangle$, since f is differentiable. The function g satisfies

$$\lim_{x,y \to x_0, x \neq y} \frac{g(x) - g(y)}{|x - y|} = 0 = \lim_{x,y \to x_0, x \neq y} \frac{|g(x) - g(y)|}{|x - y|},$$

hence for $\varepsilon = 1$, there exists a $\delta > 0$ such that $\frac{|g(x)-g(y)|}{|x-y|} < 1$ for every $x, y \in B(x_0, \delta)$, in other words:

$$|g(x) - g(y)| \leq |x - y| \quad \text{for every} \quad x, y \in B(x_0, \delta).$$

This proves that g, and consequently f, are locally Lipschitz around x_0. On the other hand g is differentiable at x_0 with $\nabla g(x_0) = 0$, hence $0 \in \partial_L g(x_0)$.

We are going to prove that there is no other vector belonging to $\partial_L g(x_0)$. If a vector $v \neq 0$ belongs to $\partial_L g(x_0)$, we have that for $\frac{|v|}{4} > 0$ there is a positive r such that $\frac{|g(x)-g(y)|}{|x-y|} < \frac{1}{4}|v|$ for every $x, y \in B(x_0, r)$. We observe now that $v \in \partial_L g(x_0)$ implies that there are a point $z \in B(x_0, r)$ and a vector $w \in \partial g(z)$ such that $|v - w| < \frac{1}{2}|v|$, then $|w| > \frac{1}{2}|v|$ necessarily. For $t > 0$ small, we have that

$$\frac{g(z + tw) - g(z) - \langle w, tw \rangle}{|tw|} = \frac{g(z + tw) - g(z)}{|tw|} - \frac{\langle w, tw \rangle}{t|w|}$$

$$< \frac{1}{4}|v| - |w| < -\frac{1}{4}|v|;$$

this implies

$$\liminf_{h\to 0} \frac{g(z+h) - g(z) - \langle w, h\rangle}{|h|} < 0,$$

which is not possible since $w \in \partial g(z)$. We conclude that $\partial_L g(x_0) = \{0\}$ and consequently $\partial_L f(x_0) = \nabla f(x_0) + \partial_L g(x_0) = \nabla f(x_0) + \{0\} = \{\nabla f(x_0)\}$. This establishes (ii).

(ii) $\Rightarrow$ (i). Let $\partial_L f(x_0) = \{v\}$. We consider the auxiliary function $g(x) = f(x) - \langle v, x\rangle$ again, we have $\partial_L g(x_0) = \{0\}$ and g locally Lipschitz around x_0. It is enough to prove that g is strictly differentiable, since it implies f strictly differentiable too. Let us see it.

We claim that for every sequence $\{x_n\}$ converging to x_0 and every sequence $\{\zeta_n\}$ such that $\zeta_n \in \partial g(x_n)$, we have that $\lim_n \zeta_n = 0$. This is true because the sequence $\{\zeta_n\}$ is bounded since g is Lipschitz near x_0, hence if it does not converge to 0 there would exist a subsequence converging to a nonzero element of $\partial_L g(x_0)$. This implies that

$$\lim_n \left[\max_{x\in B(x_0, \frac{1}{n})} \{|\zeta| : \zeta \in \partial g(x)\} \right] = 0.$$

Let us denote $\alpha_n = \max_{x\in B(x_0, \frac{1}{n})}\{|\zeta| : \zeta \in \partial g(x)\}$. The same argument that we used in Theorem 6.1, allows us to claim that g is α_n-Lipschitz in $B(x_0, \frac{1}{n})$, hence

$$\frac{|g(x) - g(y)|}{|x - y|} \le \alpha_n \quad \text{for every} \quad x, y \in B\left(x_0, \frac{1}{n}\right).$$

We deduce that $\lim_{x,y\to x_0} \frac{|g(x)-g(y)|}{|x-y|} = 0$ since $\lim_n \alpha_n = 0$, therefore g, and consequently f, are strictly differentiable at x_0.

(i) $\Rightarrow$ (iii). f is clearly locally Lipschitz near x_0. From (ii) we deduce that $\partial_L f(x_0)$ is a singleton, on the other hand $\partial f(x_0)$ is nonempty since f is differentiable, hence $\partial_L f(x_0) = \partial f(x_0)$ and f is regular. Since f is strictly differentiable if and only if $-f$ is also strictly differentiable, we have that $-f$ is regular too.

(iii) $\Rightarrow$ (ii). We have that $\partial_L f(x_0)$ and $\partial_L(-f)(x_0)$ are nonempty, by Proposition 6.2. Regularity of both f and $-f$ leads to

$$\partial f(x_0) \ne \emptyset \ne \partial(-f)(x_0),$$

but $\partial^+ f(x_0) = -\partial(-f)(x_0) \neq \emptyset$. This implies that f is differentiable at x_0 and we conclude that $\partial_L f(x_0) = \partial f(x_0)$ is a singleton.

As $-f$ is also locally Lipschitz, the same argument is valid for that function. $\square$.

The following is an in depth example of the above Theorem, making clear that there are no unnecessary conditions in it.

Example. Let $f : \mathbf{R} \to \mathbf{R}$ be defined by $f(t) = -|t|^{\frac{1}{3}}$. It is easy to see that f is uniformly continuous, but not locally Lipschitz near 0, hence it is not strictly differentiable; as a matter of fact it is easy to see it is not differentiable at 0. However, f as well as $-f$ are regular at 0, since

$$\partial_L f(0) = \partial f(0) = \emptyset \quad \text{and} \quad \partial_L(-f)(0) = \partial(-f)(0) = \mathbf{R}.$$

$\partial_L f(0)$ is not a singleton for this function, but if we consider $g(t) = f(t)$ if $t < 0$ and $g(t) = 0$ otherwise, we have an example of a function for which $\partial_L g(0) = \{0\}$ is a singleton, but that is not regular at 0 since $\partial g(0) = \emptyset$ (See Fig. 6.1).

This example points out the role of Lipschitz's property in Theorem 6.5. The next Corollary proves that strictly differentiability is a sort of pointwise C^1 condition.

COROLLARY 6.6. *Let $f : \mathbf{R}^n \to [-\infty, +\infty]$, $G \subset \text{dom} f$ be an open set. Then the following properties are equivalent:*

(i) *f is C^1 on G.*
(ii) *f is strictly differentiable on G.*
(iii) *f is locally Lipschitz and regular at every point $x \in G$ with $\partial_L f(x)$ a singleton.*

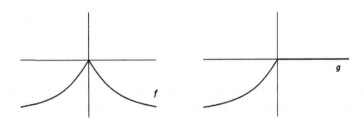

Fig. 6.1 Theorem 6.5: examples.

PROOF. Theorem 6.5 gives us immediately the equivalence between (ii) and (iii). A C^1 function on G is clearly locally Lipschitz at every $x \in G$, moreover $\partial_L f(x) = \{\nabla f(x)\} = \partial f(x)$ hence it is regular and $\partial_L f(x)$ is a singleton. This proves that (iii) follows from (i). Let us see the converse. We observe first that strict differentiability implies differentiability, that is $\partial f(x) = \{\nabla f(x)\}$ for every $x \in G$. In order to prove that f is C^1 on G it only remains to see that $\lim_n \nabla f(x_n) = \nabla f(x_0)$ when $\lim_n x_n = x_0 \in G$. But this is true because otherwise, the fact that $\{\nabla f(x_n)\}$ is bounded because f is locally Lipschitz near x_0, implies that it would have a subsequence converging to another element of $\partial_L f(x_0)$ different from $\nabla f(x_0)$, which is not possible since $\partial_L f(x_0)$ is a singleton. $\qquad \square$

The following Corollary is also an immediate consequence of Theorem 6.5.

COROLLARY 6.7. *Let $f : \mathbf{R}^n \to [-\infty, +\infty]$, $G \subset \mathrm{dom} f$ be an open set such that f is locally Lipschitz and regular at every point of G. Then f is strictly differentiable at x, provided that it is differentiable at x.*

In Chapter 1 we defined directional derivatives, another interesting concept is the one-side directional derivatives. These directional derivatives extend the idea of one-side derivatives for functions defined on $\mathbf{R}$, $f'_+(x) = \lim_{t \to 0+} \frac{f(x+t) - f(x)}{t}$ and $f'_-(x) = \lim_{t \to 0-} \frac{f(x+t) - f(x)}{t}$.

DEFINITION 6.8. For a function $f : X \to (-\infty, +\infty]$, a point $x_0 \in \mathrm{dom} f$ and a vector $v \in X$, we define the one-side directional derivative at x_0 for the direction v as

$$f'(x_0, v) = \lim_{t \searrow 0} \frac{f(x_0 + tv) - f(x_0)}{t}.$$

These one-side directional derivatives agree with the directional derivatives when they exist. In particular, if f is differentiable at x_0, we have $\langle \nabla f(x_0), v \rangle = f'(x_0, v)$.

It is interesting to relate these derivatives with the subderivative function that we introduced in Definition 4.10. Remember that this function, $df(x_0) : X \to [-\infty, +\infty]$, was defined as

$$df(x_0)(w_0) = \liminf_{t \downarrow 0, w \to w_0} \frac{f(x_0 + tw) - f(x_0)}{t}.$$

We have seen, in Proposition 6.3, that if the function f is locally Lipschitz near x_0, this subderivative has the following expression:

$$df(x_0)(w_0) = \liminf_{t \searrow 0} \frac{f(x_0 + tw_0) - f(x_0)}{t}.$$

We are focusing on locally Lipschitz functions, for such functions it holds that if the one-side directional derivatives exist, then $f'(x_0, v) = df(x_0)(v)$. From the characterization of the subdifferential that we gave in Lemma 4.11 we get the following proposition.

PROPOSITION 6.9.   *Let $f : \mathbf{R}^n \to (-\infty, +\infty]$ be a function, $x_0 \in \mathrm{dom} f$, assume that f is locally Lipschitz near x_0. If f has all the one-side derivatives at x_0, then $\zeta \in \partial f(x_0)$ if and only if $\langle \zeta, v \rangle \leq f'(x_0, v)$ for every $v \in \mathbf{R}^n$.*

6.2 THE GENERALIZED GRADIENT

We are going to introduce a more general concept of directional derivatives, by means of which we will define an extension of the gradient that moreover will allow us to study nonsmooth vector functions.

DEFINITION 6.10.   Let $f : \mathbf{R}^n \to \mathbf{R}$ be a locally Lipschitz function around x_0. For every $v \in \mathbf{R}^n$ we define the generalized directional derivative at x_0 in the direction v as

$$f^\circ(x_0, v) = \limsup_{x \to x_0, t \downarrow 0} \frac{f(x + tv) - f(x)}{t}.$$

This definition has full meaning if we consider functions defined on a Hilbert space, or even a Banach space, and many of the results are true and have similar proofs in this wider setting. However, for the sake of simplicity, we will restrict to functions defined on $\mathbf{R}^n$. It is clear that $f^\circ(x_0, v)$ is well defined, and $df(x_0)(v) \leq f^\circ(x_0, v)$, but we have more elemental properties.

PROPOSITION 6.11.   *Let $f : \mathbf{R}^n \to \mathbf{R}$ be a function, locally K-Lipschitz around x_0. Then $f^\circ(x_0, v)$ is positively homogeneous and subadditive in v. Moreover $|f^\circ(x_0, v)| \leq K|v|$.*

PROOF.   From the definition it is immediate to see that $f^\circ(x_0, \lambda v) = \lambda f^\circ(x_0, v)$ for every positive λ. In order to see the subadditivity, let us observe that

$$
f^\circ(x_0, v + w) = \limsup_{x \to x_0, t \downarrow 0} \frac{f(x + tv + tw) - f(x)}{t}
$$

$$
= \limsup_{x \to x_0, t \downarrow 0} \frac{f(x + tv + tw) - f(x + tw) + f(x + tw) - f(x)}{t}
$$

$$
\leq \limsup_{x \to x_0, t \downarrow 0} \frac{f(x + tv + tw) - f(x + tw)}{t}
$$

$$
+ \limsup_{x \to x_0, t \downarrow 0} \frac{f(x + tw) - f(x)}{t} = f^\circ(x_0, v) + f^\circ(x_0, w)
$$

since

$$
\limsup_{x \to x_0, t \downarrow 0} \frac{f(x + tv + tw) - f(x + tw)}{t} = \limsup_{y \to x_0, t \downarrow 0} \frac{f(y + tv) - f(y)}{t}
$$

via the change of variable $y = x + tw$.

Finally, let us observe that

$$
\frac{|f(x + tv) - f(x)|}{t} \leq K|v|
$$

and consequently $|f^\circ(x_0, v)| \leq K|v|$.   □

From this proposition we deduce that the generalized directional derivative satisfies

$$
f^\circ(x_0, tv + (1 - t)w) \leq f^\circ(x_0, tv) + f^\circ(x_0, (1 - t)w)
$$
$$
= t f^\circ(x_0, v) + (1 - t) f^\circ(x_0, w),
$$

in other words: it is convex in the variable v. We are now ready to define the generalized gradient.

DEFINITION 6.12. Let $f : \mathbf{R}^n \to \mathbf{R}$ be a function that we assume locally K-Lipschitz around x_0. We define the generalized gradient of f at x_0, $\overline{\nabla} f(x_0)$ as the set of vectors ζ that satisfy $\langle \zeta, v \rangle \leq f^\circ(x_0, v)$ for every $v \in \mathbf{R}^n$.

From the definition and Lemma 4.11 it is immediate to see that $\partial f(x_0) \subset \overline{\nabla} f(x_0)$, since $df(x_0)(v) \leq f^\circ(x_0, v)$.

We may consider the generalized gradient under a different scope. For a $\zeta \in \overline{\nabla} f(x_0)$ we can write $\langle \zeta, v \rangle \leq f^\circ(x_0, v) - f^\circ(x_0, 0)$ for every v, and we deduce that

$$\overline{\nabla} f(x_0) = \partial(f^\circ(x_0, .))(0)$$

since $f^\circ(x_0, .)$ is convex. We proceed to list some elementary properties of the generalized gradient.

PROPOSITION 6.13. *For a function* $f : \mathbf{R}^n \to \mathbf{R}$, *locally K-Lipschitz around x_0, the generalized gradient satisfies the following properties:*

(i) $\overline{\nabla} f(x_0)$ *is nonempty, convex and closed.*

(ii) $\overline{\nabla} f(x_0) \subset \overline{B}(0, K)$.

(iii) $\partial_L f(x_0) \subset \overline{\nabla} f(x_0)$.

PROOF. The fact that $\overline{\nabla} f(x_0)$ is the subdifferential of a convex function gives us (i) immediately. If $\zeta \in \overline{\nabla} f(x_0)$ then

$$|\langle \zeta, v \rangle| \leq |f^\circ(x_0, v)| \leq K|v| \quad \text{for every} \quad v \in \mathbf{R}^n,$$

by Proposition 6.11. Particularizing $v = \zeta$ we deduce $|\zeta|^2 \leq K|\zeta|$, which gives us (ii). Part (iii) requires some more work.

If $\zeta \in \partial_L f(x_0)$, then there are sequences $\{x_n\}$ converging to x_0 and $\{\zeta_n\}$ converging to ζ such that $\zeta_n \in \partial f(x_n)$. These ζ_n satisfy

$$\langle \zeta_n, v \rangle \leq df(x_n)(v) \leq f^\circ(x_n, v) \quad \text{for every} \quad v \in \mathbf{R}^n.$$

We claim that $f^\circ(., v)$ is *usc* at x_0, hence

$$\langle \zeta, v \rangle = \lim_n \langle \zeta_n, v \rangle \leq \limsup_n f^\circ(x_n, v) \leq f^\circ(x_0, v) \quad \text{for every} \quad v \in \mathbf{R}^n.$$

Therefore $\zeta \in \overline{\nabla} f(x_0)$. Let us prove the claim. Let x_0 and v be fixed, consider a sequence $\{x_n\}$ converging to x_0. By definition of the upper limit we have that for every n there exist $y_n \in \mathbf{R}^n$ and $t_n > 0$ such that

$$|y_n - x_n| < \frac{1}{n}, \quad t_n < \frac{1}{n}, \quad \text{and} \quad f^\circ(x_n, v) - \frac{1}{n} < \frac{f(y_n + t_n v) - f(y_n)}{t_n}.$$

Upon taking upper limits, with $n \to \infty$, we deduce $\limsup_n f^\circ(x_n, v) \leq f^\circ(x_0, v)$. □

Once we have introduced this nonsmooth gradient, we are going to derive some calculus results. We start with Fermat's Rule and the behavior under vector space operations.

PROPOSITION 6.14. *For a function $f : \mathbf{R}^n \to \mathbf{R}$, locally K-Lipschitz around x_0, the following properties hold:*

(i) *For every $\lambda \in \mathbf{R}$ we have $\overline{\nabla}(\lambda f)(x_0) = \lambda \overline{\nabla} f(x_0)$.*

(ii) $\partial^L f(x_0) \subset \overline{\nabla} f(x_0)$.

(iii) *If f has a local extreme at x_0, then $0 \in \overline{\nabla} f(x_0)$.*

(iv) *If $g : \mathbf{R}^n \to \mathbf{R}$ is also locally K-Lipschitz around x_0, then*

$$\overline{\nabla}(f + g)(x_0) \subset \overline{\nabla} f(x_0) + \overline{\nabla} g(x_0).$$

PROOF. (i) is immediate if $\lambda \geq 0$ since in this case $f^\circ(x_0, \lambda v) = \lambda f^\circ(x_0, v)$ by Proposition 6.11. In order to establish the result it is enough to prove that $\overline{\nabla}(-f)(x_0) = -\overline{\nabla} f(x_0)$, but this is true since $\zeta \in \overline{\nabla}(-f)(x_0)$ if and only if

$$\langle \zeta, v \rangle \leq (-f)^\circ(x_0, v) = f^\circ(x_0, -v) \quad \text{for every} \quad v \in \mathbf{R}^n.$$

And this is equivalent to $\langle -\zeta, v \rangle \leq f^\circ(x_0, v)$ for every $v \in \mathbf{R}^n$, which holds if and only if $-\zeta \in \overline{\nabla} f(x_0)$. The equality $(-f)^\circ(x_0, v) = f^\circ(x_0, -v)$ that we claimed is an easy consequence of the definition via a change of variable, let us see it:

$$(-f)^\circ(x_0, v) = \limsup_{x \to x_0, t \downarrow 0} \frac{-f(x + tv) + f(x)}{t}$$

$$= \limsup_{y \to x_0, t \downarrow 0} \frac{f(y - tv) - f(y)}{t} = f^\circ(x_0, -v).$$

For (ii) it is enough to observe that

$$\partial^L f(x_0) = -\partial_L(-f)(x_0) \subset -\overline{\nabla}(-f)(x_0) = -\left[-\overline{\nabla} f(x_0)\right] = \overline{\nabla} f(x_0).$$

(iii) is an immediate consequence of (ii) since if f has a minimum, $0 \in \partial f(x_0)$, while if it has a maximum $0 \in \partial^+ f(x_0)$.

To prove (iv), we claim that $(f + g)^\circ(x_0, v) \leq f^\circ(x_0, v) + g^\circ(x_0, v)$ which implies

$$\overline{\nabla}(f + g)(x_0) = \partial\big((f + g)^\circ(x_0, .)\big)(0) \subset \partial\big(f^\circ(x_0, .) + g^\circ(x_0, .)\big)(0)$$
$$\subset \partial\big(f^\circ(x_0, .)\big)(0) + \partial\big(g^\circ(x_0, .)\big)(0) = \overline{\nabla} f(x_0) + \overline{\nabla} g(x_0).$$

The first inclusion is a consequence of the characterization of the subdifferential in Theorem 4.1, and the fact that $(f + g)^\circ(x_0, v) \leq f^\circ(x_0, v) + g^\circ(x_0, v)$ and $(f + g)^\circ(x_0, 0) = f^\circ(x_0, 0) + g^\circ(x_0, 0)$, while the second one, that is actually an equality, is the elementary Sum Rule for convex functions, see Propositions 5.15 and 5.17 for instance.

Now we verify that the claim holds:

$$(f + g)^\circ(x_0, v) = \limsup_{x \to x_0, t \downarrow 0} \frac{f(x + tv) + g(x + tv) - f(x) - g(x)}{t}$$

$$\leq \limsup_{x \to x_0, t \downarrow 0} \frac{f(x + tv) - f(x)}{t} + \limsup_{x \to x_0, t \downarrow 0} \frac{g(x + tv) - g(x)}{t}$$

$$= f^\circ(x_0, v) + g^\circ(x_0, v). \qquad \Box$$

The generalized gradient provides us with an exact Mean Value Theorem. It is enough to observe that $\partial^L f(x_0) \subset \overline{\nabla} f(x_0)$ and reformulate Proposition 5.21, to obtain

COROLLARY 6.15. *Let $f : \mathbf{R}^n \to \mathbf{R}$ be locally Lipschitz around a line segment $[x, y]$, then there are a point $z_0 \in [x, y]$ and a vector $\zeta_0 \in \overline{\nabla} f(z_0)$ such that*

$$f(y) - f(x) = \langle \zeta_0, y - x \rangle$$

What is the generalized gradient for a C^1 function? As expected it agrees with the gradient.

PROPOSITION 6.16. *Let $f : \mathbf{R}^n \to \mathbf{R}$, $x_0 \in \mathbf{R}^n$, let us assume that f is C^1 in a neighborhood of x_0, then $\overline{\nabla} f(x_0) = \{\nabla f(x_0)\}$.*

PROOF.   From the smooth Mean Value Theorem we deduce that $\frac{f(x+tv)-f(x)}{t} = \langle \nabla f(z), v \rangle$ with $z \in [x, x + tv]$. Taking upper limits we have $f^\circ(x_0, v) = \langle \nabla f(x_0), v \rangle$ and consequently $\zeta \in \overline{\nabla} f(x_0)$ if and only if $\langle \zeta, v \rangle \leq \langle \nabla f(x_0), v \rangle$ for every $v \in \mathbf{R}^n$, which is equivalent to $\zeta = \nabla f(x_0)$. $\qquad \Box$

We have seen above that $\partial_L f(x_0) \subset \overline{\nabla} f(x_0)$. This inclusion is clearly strict, as the following example shows.

Example.   Let us consider the one variable function $f(t) = -|t|$. We proved above that it satisfies $\partial_L f(0) = \{-1, 1\}$, when $[-1, 1] = \partial^+ f(0) \subset$

$\overline{\nabla} f(0)$; in fact we have that $\overline{\nabla} f(0) = [-1, 1]$. In order to see this we must check the value of $f°(0, 1)$ and $f°(0, -1)$.

$$f°(0, 1) = \limsup_{x \to 0, t \downarrow 0} \frac{-|x + t| + |x|}{t} \leq \limsup_{x \to x_0, t \downarrow 0} \frac{t}{t} = 1$$

and the value 1 is attained if we restrict the upper limit to $x = -t$. We conclude that $f°(0, 1) = 1$. Analogously we have $f°(0, -1) = 1$ too. Hence $r \in \overline{\nabla} f(0)$ if and only if $r \cdot 1 \leq f°(0, 1) = 1$ and $r \cdot (-1) \leq f°(0, -1) = 1$, that is $|r| \leq 1$.

Later we will see when the equality holds, but before that we establish the following interesting theorem.

THEOREM 6.17. *Let* $f : \mathbf{R}^n \to \mathbf{R}$ *be a function, locally Lipschitz around* x_0, *then we have*

$$\overline{\nabla} f(x_0) = \overline{co}(\partial_L f(x_0)).$$

PROOF. From $\partial_L f(x_0) \subset \overline{\nabla} f(x_0)$ we deduce $\overline{co}(\partial_L f(x_0)) \subset \overline{\nabla} f(x_0)$ since $\overline{\nabla} f(x_0)$ is closed and convex by Proposition 6.13. In order to see the opposite inclusion, we suppose that there is a vector $\xi \in \overline{\nabla} f(x_0)$ such that $\xi \notin \overline{co}(\partial_L f(x_0))$, and let $r > 0$ be such that

$$B(\xi, r) \cap \overline{co}(\partial_L f(x_0)) = \emptyset.$$

We invoke Minkowski's Separation Theorem with $C_1 = B(\xi, r)$ and $C_2 = \overline{co}(\partial_L f(x_0))$, that satisfy $0 \notin C_1 \setminus C_2 = int(C_1 \setminus C_2)$, and obtain that there is a vector $v \in \mathbf{R}^n$ such that $\langle \zeta, v \rangle < \langle \xi, v \rangle$ for every $\zeta \in C_2$. We claim that there is a vector $\zeta_0 \in \partial_L f(x_0)$ such that $f°(x_0, v) \leq \langle \zeta_0, v \rangle$ and consequently

$$f°(x_0, v) \leq \langle \zeta_0, v \rangle < \langle \xi, v \rangle \leq f°(x_0, v)$$

since $\xi \in \overline{\nabla} f(x_0)$, which gives us a contradiction that proves the required inclusion. Let us prove the claim.

Let the sequences $\{x_n\}$ converging to x_0 and $t_n \searrow 0$ define $f°(x_0, v)$, that is

$$f°(x_0, v) = \lim_n \frac{f(x_n + t_n v) - f(x_n)}{t_n}.$$

For every n, Proposition 5.19 implies that there exist a point $z_n \in [x_n, x_n + t_n v] + \frac{t_n}{n} B$, and a vector $\zeta_n \in \partial f(z_n)$ such that $f(x_n + t_n v) - f(x_n) \leq \langle \zeta_n, t_n v \rangle + \frac{t_n}{n}$. Clearly $\lim_n z_n = x_0$, and we may assume that $\{\zeta_n\}$ also converges since it is bounded by the Lipschitzness of f near x_0, the vector $\zeta_0 = \lim_n \zeta_n$ belongs to $\partial_L f(x_0)$. Therefore the claim follows from

$$f^\circ(x_0, v) = \lim_n \frac{f(x_n + t_n v) - f(x_n)}{t_n} \leq \limsup_n \frac{\langle \zeta_n, t_n v \rangle + \frac{t_n}{n}}{t_n}$$

$$= \lim_n \left(\langle \zeta_n, v \rangle + \frac{1}{n} \right) = \langle \zeta_0, v \rangle. \qquad \square$$

This theorem has some interesting consequences. In the setting of locally Lipschitz functions, let us assume that f is regular at x_0, we have that $\partial_L f(x_0)$ is closed and convex since it agrees with $\partial f(x_0)$ hence $\overline{co}(\partial_L f(x_0)) = \partial_L f(x_0)$ and consequently

$$\partial f(x_0) = \partial_L f(x_0) = \overline{\nabla} f(x_0).$$

On the other hand, we know that $df(x_0)(v) \leq f^\circ(x_0, v)$, but in the proof of the above theorem we have seen that for every v there is a $\zeta_0 \in \partial_L f(x_0) = \partial f(x_0)$ such that $f^\circ(x_0, v) \leq \langle \zeta_0, v \rangle$. Joining all the inequalities we have:

$$df(x_0)(v) \leq f^\circ(x_0, v) \leq \langle \zeta_0, v \rangle \leq df(x_0)(v),$$

hence

$$df(x_0)(v) = \liminf_{t \searrow 0} \frac{f(x_0 + tv) - f(x_0)}{t}$$

$$\leq \limsup_{t \searrow 0} \frac{f(x_0 + tv) - f(x_0)}{t} \leq f^\circ(x_0, v) = df(x_0)(v).$$

We have proved that f has one-side directional derivatives and $f^\circ(x_0, v) = f'(x_0, v)$ for every v. The converse is also true.

PROPOSITION 6.18. *Let $f : \mathbf{R}^n \to \mathbf{R}$, assume that it is locally Lipschitz near x_0, then f is regular if and only if it has one-side directional derivatives and $f'(x_0, v) = f^\circ(x_0, v)$ for every v. Moreover we have*

$$\partial f(x_0) = \partial_L f(x_0) = \overline{\nabla} f(x_0).$$

PROOF. We only have to prove that the existence of one-side directional derivatives plus the equality $f'(x_0, v) = f°(x_0, v)$ guarantee the regularity of f. The existence of one-side directional derivatives for Lipschitz functions implies that $df(x_0)(v) = f'(x_0, v)$ for every $v \in \mathbf{R}^n$. This equality together with our assumption leads to $df(x_0)(v) = f°(x_0, v)$ therefore $\partial f(x_0) = \overline{\nabla} f(x_0)$, and necessarily $\partial f(x_0) = \partial_L f(x_0)$. In other words f is regular. $\square$

Of course, from this proposition an exact Sum Rule for regular functions follows, that is:

COROLLARY 6.19. *Let $f, g : \mathbf{R}^n \to \mathbf{R}$ be two functions, let us assume that they are locally Lipschitz near x_0 and regular at that point. Then*

$$\overline{\nabla}(f + g)(x_0) = \overline{\nabla} f(x_0) + \overline{\nabla} g(x_0).$$

Another immediate consequence of Theorem 6.17 is the following Chain Rule inclusion.

COROLLARY 6.20. *Let $F : \mathbf{R}^n \to \mathbf{R}^m$ be a C^1 function and $x_0 \in \mathbf{R}^m$, assume that the function $f : \mathbf{R}^m \to \mathbf{R}$ is locally Lipschitz near $F(x_0)$. Then*

$$\overline{\nabla}(f \circ F)(x_0) \subset DF(x_0)^* \big(\overline{\nabla} f(F(x_0)) \big).$$

PROOF. It is enough to observe that Proposition 5.11 implies

$$\partial_L(f \circ F)(x_0) \subset DF(x_0)^* \big(\partial_L f(F(x_0)) \big) \subset DF(x_0)^* \big(\overline{\nabla} f(F(x_0)) \big).$$

This last set is convex and closed since it is a linear transformation of $\overline{\nabla} f(F(x_0))$ which is convex and closed. Hence $\overline{\nabla}(f \circ F)(x_0) = \overline{co}(\partial_L (f \circ F)(x_0))$ is also contained in $DF(x_0)^* \big(\overline{\nabla} f(F(x_0)) \big)$. $\square$

Proposition 6.18 allows us to translate the exact Chain Rule that we proved in Proposition 5.16 when the function f is regular, namely:

COROLLARY 6.21. *Let $F : \mathbf{R}^n \to \mathbf{R}^m$ be a C^1 function and $x_0 \in \mathbf{R}^m$, assume that the function $f : \mathbf{R}^m \to \mathbf{R}$ is locally Lipschitz near $F(x_0)$ and regular at $F(x_0)$. Then*

$$\overline{\nabla}(f \circ F)(x_0) = DF(x_0)^* \big(\overline{\nabla} f(F(x_0)) \big).$$

PROOF. It is enough to invoke Propositions 6.18 and 5.16, and observe that, as we proved in Proposition 5.18, $f \circ F$ is regular at x_0. $\square$

In order to finish this section, we are going to show that the generalized gradient can be approximated by regular gradients. But a problem arises: given a point x is there a sequence $\{x_n\}$ converging to x such that $\nabla f(x_n)$ exists for every n? The answer is yes, but we can say much more by means of Rademacher's Theorem.

THEOREM 6.22 (*Rademacher's Theorem*).   *Every Lipschitz function $f : \mathbf{R}^n \to \mathbf{R}^m$ is differentiable almost everywhere.*

Almost everywhere means that the property holds for every point except those in a negligible set. The proof of this theorem is out of the scope of this book, but we are going to invoke it to deduce that for every $x \in \mathbf{R}^n$ there is a sequence such that $\lim_n x_n = x$ and such that $\nabla f(x_n)$ exists for every n; in fact there are "plenty" sequences like that. This is an immediate consequence of the fact that open balls have positive measure and consequently for every positive ε there are many points $z \in B(x, \varepsilon)$ such that $\nabla f(z)$ exists.

An elementary property of a negligible subset of $\mathbf{R}^n$, N, is that for every $v \in \mathbf{R}^n$ there is a negligible set A_v such that the sets

$$N_x = \{x + tv : t \in \mathbf{R}\} \cap N$$

are negligible, as one-dimensional sets, for every $x \in \mathbf{R}^n \setminus A_v$. We will omit the proof of this fact, which follows from Fubini's Theorem. Once established this remark, in order to prove the following Theorem, we require a technical Lemma.

LEMMA 6.23.   *Let $f : \mathbf{R}^n \to \mathbf{R}$ be a K-Lipschitz function and $v \in \mathbf{R}^n$, we have*

$$f^\circ(x_0, v) \le \limsup_{y \to x_0} \langle \nabla f(y), v \rangle$$

where the upper limit is taken over points of differentiability y of f.

PROOF.   Rademacher's Theorem asserts that there is a negligible set N such that f is differentiable at every $x \notin N$. Fix a vector $v \in \mathbf{R}^n$, from the remark above we know that almost all N_x are negligible. On the other hand there exist sequences $\{x_n\}$ converging to x_0 and $\{t_n\}$ decreasing to 0 such that

$$f^\circ(x_0, v) = \lim_n \frac{f(x_n + t_n v) - f(x_n)}{t_n}.$$

We may assume that all the sets N_{x_n} are negligible since otherwise we would replace x_n by $\tilde{x}_n$ to be such that $|x_n - \tilde{x}_n| < \frac{t_n}{2nK}$, with $N_{\tilde{x}_n}$ negligible. Once we have justified our assumption we estimate the quotients $\frac{f(x_n + t_n v) - f(x_n)}{t_n}$.

The function $\varphi_n : \mathbf{R} \to \mathbf{R}$ defined by $\varphi_n(t) = f(x_n + tv)$ is differentiable almost everywhere with derivative $\varphi_n'(t) = \langle \nabla f(x_n + tv), v \rangle$, hence

$$\frac{f(x_n + t_n v) - f(x_n)}{t_n} = \frac{\varphi_n(t_n) - \varphi_n(0)}{t_n}$$

$$= \frac{1}{t_n} \int_0^{t_n} \varphi_n'(t)dt = \frac{1}{t_n} \int_0^{t_n} \langle \nabla f(x_n + tv), v \rangle dt$$

$$\leq \sup\{\langle \nabla f(x_n + tv), v \rangle : t \in [0, t_n], x_n + tv \notin N\}.$$

If we denote $r_n = |x_n + t_n v - x_0|$, we have

$$\frac{f(x_n + t_n v) - f(x_n)}{t_n} \leq \sup\{\langle \nabla f(y), v \rangle : y \in B(x_0, r_n) \setminus N\},$$

and by taking limits we conclude

$$f^\circ(x_0, v) \leq \limsup_{y \notin N, y \to x_0} \langle \nabla f(y), v \rangle. \qquad \square$$

Now, we may prove the following theorem.

THEOREM 6.24.   *Let $x_0 \in \mathbf{R}^n$ and $f : \mathbf{R}^n \to \mathbf{R}$ be a Lipschitz function near x_0. Then*

$$\overline{\nabla} f(x_0) = co\{\lim_n \nabla f(x_n) : f \text{ is differentiable at } x_n, \quad \lim_n x_n = x_0\}.$$

PROOF.   Let us observe first that

$$A = \{\lim_n \nabla f(x_n) : f \text{ is differentiable at } x_n, \quad \lim_n x_n = x_0\} \subset \partial_L f(x_0)$$

since $\nabla f(x_n) \in \partial f(x_n)$ when f is differentiable at x_n. This implies that its convex hull is contained in $\overline{co}\partial_L f(x_0) = \overline{\nabla} f(x_0)$.

For the other inclusion, we observe first that A is closed since

$$v = \lim_k \left[\lim_n \nabla f(x_n^{(k)})\right] \quad \text{with} \quad \lim_n x_n^{(k)} = x_0 \quad \text{for every } k$$

implies $v = \lim_n \nabla f(x_n^{(n)})$ and $\lim_n x_n^{(n)} = x_0$. On the other hand A is also bounded because of the Lipschitz character of f, hence A is compact, and consequently so is its convex hull. We are going to suppose that there is a vector $v_0 \in \overline{\nabla} f(x_0)$ such that $v_0 \notin coA$ and we will arrive at a contradiction. Such vector v_0 satisfies, by Minkowski's Separation Theorem, that there is another vector v and a real number α such that $\langle v, v_0 \rangle > \alpha$ and $\langle v, w \rangle \leq \alpha$ for every $w \in A$. $v_0 \in \overline{\nabla} f(x_0)$ implies

$$\alpha < \langle v, v_0 \rangle \leq f^\circ(x_0, v).$$

By Lemma 6.23, we have

$$f^\circ(x_0, v) \leq \limsup_{y \to x_0} \langle \nabla f(y), v \rangle$$

where the upper limit is taken over points y such that $\nabla f(y)$ exists. From $\langle v, w \rangle \leq \alpha$ for every $w \in A$ we deduce

$$\limsup_{y \to x_0} \langle v, \nabla f(y) \rangle \leq \alpha,$$

which leads to the contradiction $\alpha < \alpha$.   □

6.3 GENERALIZED JACOBIAN

We have not defined nonsmooth concepts for vector valued functions, there are several reasons for this: the role of the inequalities in the theory presented so far and the non existence of epigraphs among others. However Theorem 6.24 together with Rademacher's Theorem suggests a way to define a generalized Jacobian. In the next section we will introduce a more geometrical concept and study the relation between both ideas. We start with a definition.

DEFINITION 6.25. Let $F : \mathbf{R}^n \to \mathbf{R}^m$ be a Lipschitz function. We define the Generalized Jacobian of F at a point x_0 as

$$\overline{J} F(x_0) = co\{\lim_n J F(x_n) : F \text{ is differentiable at } x_n, \quad \lim_n x_n = x_0\},$$

where $J F(x)$ denotes the Jacobian matrix of F at x.

Limits and linear operations in the definition above are the usual ones in the vector space of $m \times n$ matrices, $M_{m \times n}$. It is immediate to realize that $\overline{J} F(x_0)$ is a compact convex subset of $M_{m \times n}$. The next proposition is an easy consequence of Theorem 6.24.

PROPOSITION 6.26. *Let* $F : \mathbf{R}^n \to \mathbf{R}^m$ *be a Lipschitz function, and* $v \in \mathbf{R}^m$. *Let us denote* $f(x) = \langle F(x), v \rangle$. *Then*

$$\overline{\nabla} f(x_0) = \{ Mv : M \in \overline{J} F(x_0) \}$$

for every $x_0 \in \mathbf{R}^n$.

PROOF. A matrix $M_0 \in \overline{J} F(x_0)$ is a convex combination of matrices $M = \lim_n J F(x_n)$ with $\{x_n\}$ converging to x_0. We have that for such matrices M

$$Mv = \lim_n J F(x_n)v = \lim_n \nabla f(x_n)$$

by the smooth chain rule, and consequently $M_0 v$ is a convex combination of such limits of gradients. Hence $M_0 v \in \overline{\nabla} f(x_0)$ by Theorem 6.24. The opposite inclusion is similar. $\square$

We may consider $\overline{J} F$ as a function that assigns points to subsets of matrices, that is: a set valued function from $\mathbf{R}^n$ to $M_{m \times n}$. We will develop elementary continuity properties in the problem section below; the next proposition asserts, in the terminology that we will introduce, that $\overline{J} F$ is outer semicontinuous.

PROPOSITION 6.27. *Let* $F : \mathbf{R}^n \to \mathbf{R}^m$ *be a Lipschitz function,* $x_0 \in \mathbf{R}^n$. *For every* $\varepsilon > 0$ *there exists a positive* r, *such that*

$$\overline{J} F(x) \subset \overline{J} F(x_0) + \varepsilon B$$

for every $x \in B(x_0, r)$.

PROOF. Let us observe that the norm that we consider on $M_{m \times n}$ does not matter since all of them are equivalent. On the other hand it is enough to prove the inclusion column by column, hence Proposition 6.26 allows us to assume without loss of generality that $m = 1$. In other words: we will prove the proposition for the generalized gradient. Let us suppose that there exist a sequence $\{x_n\}$ converging to x_0, and vectors $\zeta_n \in \overline{\nabla} F(x_n)$ such that $\zeta_n \notin \overline{\nabla} F(x_0) + \varepsilon B$. This set is convex and open hence, by Minkowski's Separation Theorem, there are non zero vectors v_n such that

$$\langle \zeta_n, v_n \rangle \geq \sup\{\langle \zeta, v_n \rangle : \zeta \in \overline{\nabla} F(x_0) + \varepsilon B\}$$
$$= \sup\{\langle \zeta, v_n \rangle : \zeta \in \overline{\nabla} F(x_0)\} + \varepsilon |v_n| = F^0(x_0, v_n) + \varepsilon |v_n|, \quad (6.1)$$

dividing by $|v_n|$ everywhere, we may assume that the vectors v_n have norm 1. Passing to a subsequence if necessary, we may assume also that the sequences $\{v_n\}$ and $\{\zeta_n\}$ converge to v and ζ respectively. Taking upper limits in (6.1) we get

$$\langle \zeta, v \rangle \geq F^0(x_0, v) + \varepsilon$$

since $F^0(x_0, .)$ is usc.

On the other hand, we have

$$F^0(x_0, v) \geq \limsup_n F^0(x_n, v) \geq \limsup_n \langle \zeta_n, v \rangle = \langle \zeta, v \rangle$$

since $F^0(., v)$ is usc and $\zeta_n \in \overline{\nabla} F(x_n)$. We have arrived at the contradiction $\langle \zeta, v \rangle \geq \langle \zeta, v \rangle + \varepsilon$.   □

Our next goal is to present an Inverse Function Theorem. The condition that we will require is non singularity of the matrices that belong to the Generalized Jacobian. Let us precise it:

DEFINITION 6.28.   Let $F : \mathbf{R}^n \to \mathbf{R}^n$ be a Lipschitz function, $x_0 \in \mathbf{R}^n$. We say that $\overline{J}F(x_0)$ is nonsingular if all the matrices belonging to $\overline{J}F(x_0)$ are nonsingular.

The following Lemma highlights the meaning of the non singular generalized Jacobian.

LEMMA 6.29.   *Let $F : \mathbf{R}^n \to \mathbf{R}^n$ be a Lipschitz function, $x_0 \in \mathbf{R}^n$. Assume that $\overline{J}F(x_0)$ is nonsingular. Then there are positive numbers r and δ such that for every unit vector $v \in \mathbf{R}^n$ there is a unit vector $w \in \mathbf{R}^n$ such that $\langle w, Mv \rangle \geq \delta$ for every $M \in \overline{J}F(x)$ with $x \in B(x_0, r)$.*

PROOF.   Let us consider the subset of $\mathbf{R}^n$

$$S = \{Mv : M \in \overline{J}F(x_0), |v| = 1\}.$$

S is clearly compact and $0 \notin S$, hence there exists a positive δ such that $|z| \geq 2\delta$ for every $z \in S$. From Proposition 6.27 there exists a positive r such that $\overline{J}F(x) \subset \overline{J}F(x_0) + \delta B$ for every $x \in B(x_0, r)$, therefore a matrix $M \in \overline{J}F(x)$, for $x \in B(x_0, r)$, can be written as $M = M_0 + P$ with $M_0 \in \overline{J}F(x_0)$ and $||P|| < \delta$. Given a unit vector v,

$$|Mv| = |M_0 v + Pv| \geq |M_0 v| - |Pv| \geq 2\delta - |Pv| \geq \delta,$$

hence there exists a unit vector w such that $\langle w, Mv \rangle = |Mv| \geq \delta$.   □

Let us observe that the converse is trivially true since if the inequality holds for every $M \in \overline{J} F(x)$, it holds for $M \in \overline{J} F(x_0)$ in particular. But $\langle w, Mv \rangle \geq \delta$ for a w implies that $Mv \neq 0$ for every v, that is M is nonsingular. On the other hand, for non C^1 differentiable functions, the result is no longer true if we replace generalized Jacobian by Jacobian, as the following example shows.

Example. The function $f : \mathbf{R} \to \mathbf{R}$ defined by $f(t) = t + t^2 \sin \frac{1}{t}$ for $t \neq 0$ and $f(0) = 0$, is Lipschitz, differentiable, with $f'(0) = 1$, but it does not satisfy the Lemma, because it is not C^1.

Now we are ready to prove the Lipschitz Inverse Function Theorem.

THEOREM 6.30. *Let $F : \mathbf{R}^n \to \mathbf{R}^n$ be a Lipschitz function, $x_0 \in \mathbf{R}^n$. Assume that $\overline{J} F(x_0)$ is nonsingular. Then there exist W and V neighborhoods of $F(x_0)$ and x_0 respectively, and a Lipschitz function $G : W \to V$ such that $F(G(y)) = y$ for every $y \in W$ and $G(F(x)) = x$ for every $x \in V$.*

PROOF. Let δ and r be as in Lemma 6.29, let $V = B(x_0, r)$. We are going to prove that

$$|F(x) - F(y)| \geq \delta |x - y| \quad \text{for every} \quad x, y \in V. \tag{6.2}$$

We fix $x \neq y, x, y \in V$. Let us denote $v = \frac{y-x}{|y-x|}$ and $t_0 = |y - x|$. The set N of points where F fails to be differentiable meets the line $\{z + tv : t \in \mathbf{R}\}$ in a one-dimensional negligible set for almost every $z \in B(x_0, r)$, considering only those z close enough to x for which $[z, z + t_0 v] \subset V$ holds. The function $\varphi : [0, t_0] \to \mathbf{R}^n$ defined as $\varphi(t) = F(z + tv)$ is differentiable almost everywhere with derivative $\varphi'(t) = J F(z + tv)v$. Thus

$$F(z + t_0 v) - F(z) = \int_0^{t_0} J F(z + tv) v \, dt.$$

If w is the vector provided by Lemma 6.29, we have

$$\langle w, F(z + t_0 v) - F(z) \rangle = \int_0^{t_0} \langle w, J F(z + tv) v \rangle dt \geq \int_0^{t_0} \delta \, dt = t_0 \delta,$$

hence

$$|F(z + y - x) - F(x)| \geq \delta |y - x|$$

letting $z \to x$, we obtain (6.2), since F is continuous. Once established this formula, we deduce that F is one to one from V onto $W = F(V)$. $G = F^{-1}$ satisfies

$$\left| G(y) - G(x) \right| \leq \frac{1}{\delta} \left| F(G(y)) - F(G(x)) \right| = \frac{1}{\delta} \left| y - x \right|,$$

hence it is Lipschitz. A posteriori we know that W is open, because F is an homeomorphism.   □

6.4 GRAPHICAL DERIVATIVE

As we observed in the preceding section, the lack of epigraphs is a handicap if we are dealing with vector functions. However, when comparing to the smooth case, the graph is enough in order to define differentiability. In this section we will present this approach.

DEFINITION 6.31. Let $F : \mathbf{R}^n \to \mathbf{R}^m$ be a continuous function and $x_0 \in \mathbf{R}^n$ a point. The graphical derivative of F at x_0 is the set valued function $DF(x_0) : \mathbf{R}^n \rightrightarrows \mathbf{R}^m$ defined by

$$y \in DF(x_0)(x) \quad \text{if and only if} \quad (x, y) \in T_{GraphF}(x_0, F(x_0)).$$

Similarly, we define the graphical coderivative $D^*F(x_0) : \mathbf{R}^m \rightrightarrows \mathbf{R}^n$ by

$$x \in D^*F(x_0)(y) \quad \text{if and only if} \quad (x, -y) \in N_{GraphF}(x_0, F(x_0)).$$

These definitions make sense since the continuity of F guarantees that $GraphF$ is closed and consequently we may define the tangent and the general normal. On the other hand it is not difficult to see that for a C^1 function the graphical derivative agrees with the differential, while the coderivative is the adjoint to the differential, therefore the notation is consistent. A last commentary on the definition: usually these concepts are defined for set valued functions, however for the sake of simplicity, we will deal with the single valued case only.

Let us observe that we may read the definition of graphical derivative as

$$Graph(DF(x_0)) = T_{GraphF}(x_0, F(x_0)).$$

From Proposition 4.14 we know that the graph of the graphical derivative is a closed cone, this implies in particular that $DF(x_0)$ is positively homogeneous, that is:

$$y \in DF(x_0)(x) \Rightarrow \lambda y \in DF(x_0)(\lambda x)$$

for every positive λ.

The next proposition is no more than an elementary translation of the definition of the tangent cone, but it is useful and interesting because it presents an analytical characterization of the graphical derivative.

PROPOSITION 6.32. *Let* $F : \mathbf{R}^n \to \mathbf{R}^m$ *be a continuous function,* $x_0 \in \mathbf{R}^n$. *For every* $h \in \mathbf{R}^n$ *we have that* $v \in DF(x_0)(h)$ *if and only if there are sequences* $\{h_n\}$ *converging to* h *and* $t_n \searrow 0$ *such that*

$$v = \lim_n \frac{F(x_0 + t_n h_n) - F(x_0)}{t_n}.$$

This proposition implies in particular that the one-side directional derivative at x_0 for the direction h, $f'(x_0, h)$, belongs to the graphical derivative $Df(x_0)(h)$ for continuous scalar functions f, but we may extend this result to the vectorial case easily.

DEFINITION 6.33. For a continuous function $F : \mathbf{R}^n \to \mathbf{R}^m$, a point $x_0 \in \mathbf{R}^n$ and a vector $h \in \mathbf{R}^n$, we define the one-side directional derivative at x_0 for the direction h as

$$F'(x_0, h) = \lim_{t \searrow 0} \frac{F(x_0 + th) - F(x_0)}{t}$$

whenever the limit exists.

The following result is immediate.

PROPOSITION 6.34. *Let* $F : \mathbf{R}^n \to \mathbf{R}^m$ *be a continuous function,* $x_0 \in \mathbf{R}^n$ *a point and* $h \in \mathbf{R}^n$ *a vector. Then*

$$F'(x_0, h) \in DF(x_0)(h)$$

provided that the one-side directional derivative exists.

It is not difficult to find functions F such that $DF(x_0)(h) = \emptyset$ for some necessarily nonzero h. However, for Lipschitz functions, we have the following result, which is an immediate consequence of Proposition 6.32, and compactness of closed bounded sets.

PROPOSITION 6.35. *Let $F : \mathbf{R}^n \to \mathbf{R}^m$ be a Lipschitz function, $x_0 \in \mathbf{R}^n$ a point and $h \in \mathbf{R}^n$ a vector. Then $DF(x_0)(h) \neq \emptyset$.*

The following proposition highlights the gradient character of the coderivative.

PROPOSITION 6.36. *Let $f : \mathbf{R}^n \to \mathbf{R}$ be a continuous function and $x_0 \in \mathbf{R}^n$ a point. Then*

$$\partial_L^- f(x_0) \cup \partial_L^+ f(x_0) \subset D^* f(x_0)(1).$$

PROOF.

$$\partial_L^- f(x_0) = \{\zeta \in \mathbf{R}^n : (\zeta, -1) \in N_{epi f}(x_0, f(x_0))\}$$
$$\subset \{\zeta \in \mathbf{R}^n : (\zeta, -1) \in N_{Graph f}(x_0, f(x_0))\} = D^* f(x_0)(1).$$

On the other hand

$$\partial_L^+ f(x_0) = -\partial_L^-(-f)(x_0) \subset -D^*(-f)(x_0)(1) = D^* f(x_0)(1). \qquad \square$$

For Lipschitz scalar functions we deduce, taking convex hulls, that

$$\overline{\nabla} f(x_0) \subset co\big(D^* f(x_0)(1)\big).$$

However we will present a more general result later on.

Now, we return to the vector case. The proposition that we are going to show will allow us to extend results from scalar to vector functions.

PROPOSITION 6.37. *Let $F : \mathbf{R}^n \to \mathbf{R}^m$ be a K-Lipschitz function, $x_0 \in \mathbf{R}^n$ a point, and a vector $v \in \mathbf{R}^m$. Then*

$$D^* F(x_0)(v) = \partial_L\big(\langle v, F(.)\rangle\big)(x_0).$$

PROOF. Let $h \in \mathbf{R}^n$, from the definition we have that $h \in D^* F(x_0)(v)$ if and only if $(h, -v) \in N_{Graph F}(x_0, F(x_0))$, which is equivalent to $(h, -v) = \lim_n (h_n, -v_n)$ with

$$(h_n, -v_n) \in \hat{N}_{Graph F}(x_n, F(x_n))$$

for a sequence $\{x_n\}$ converging to x_0. From 4.4 this is equivalent to

$$\langle h_n, x - x_n\rangle - \langle v_n, F(x) - F(x_n)\rangle$$
$$\leq o(|(x, F(x)) - (x_n, F(x_n))|) = o(|x - x_n|) \qquad (6.3)$$

for every x, where the last equality is a consequence of the fact that F is Lipschitz. We may write (6.3) equivalently as

$$\langle v_n, F(x)\rangle \geq \langle v_n, F(x_n)\rangle + \langle h_n, x - x_n\rangle + o(|x - x_n|), \qquad (6.4)$$

which implies that

$$\begin{aligned}
h_n \in \partial\big(\langle v_n, F(.)\rangle\big)(x_n) &= \partial\big(\langle v, F(.)\rangle + \langle v_n - v, F(.)\rangle\big)(x_n)\\
&\subset \partial_L\big(\langle v, F(.)\rangle + \langle v_n - v, F(.)\rangle\big)(x_n)\\
&\subset \partial_L\big(\langle v, F(.)\rangle\big)(x_n) + \partial_L\big(\langle v_n - v, F(.)\rangle\big)(x_n)
\end{aligned}$$

by Proposition 5.4. Hence $h_n = \tilde{h}_n + \bar{h}_n$ where $\tilde{h}_n \in \partial_L\big(\langle v, F(.)\rangle\big)(x_n)$ and $\bar{h}_n \in \partial_L\big(\langle v_n - v, F(.)\rangle\big)(x_n)$ which is a $K|v_n - v|$-Lipschitz function. We have that $\lim_n \tilde{h}_n = \lim_n h_n = h$ since $|\bar{h}_n| \leq K|v - v_n|$. We conclude that $h \in \partial_L\big(\langle v, F(.)\rangle\big)(x_0)$.

Conversely, if $h \in \partial_L\big(\langle v, F(.)\rangle\big)(x_0)$ there exist sequences $\{x_n\}$ and $\{h_n\}$ converging to x_0 and h respectively such that $h_n \in \partial\big(\langle v, F(.)\rangle\big)(x_n)$. Hence

$$\langle v, F(x)\rangle \geq \langle v, F(x_n)\rangle + \langle h_n, x - x_n\rangle + o(|x - x_n|),$$

which, as we observe above, is equivalent to

$$(h_n, -v) \in \hat{N}_{GraphF}(x_n, F(x_n)).$$

This implies

$$(h, -v) \in N_{GraphF}(x_0, F(x_0)). \qquad \square$$

The following Corollary is an immediate consequence of Proposition 6.37 for $v = 1$ and Theorem 6.17.

COROLLARY 6.38. *Let $f : \mathbf{R}^n \to \mathbf{R}$ be a Lipschitz function, $x_0 \in \mathbf{R}^n$. Then*

$$\overline{\nabla} f(x_0) = co\big(D^* f(x_0)(1)\big).$$

For Lipschitz functions the graphical derivative has a simpler expression.

PROPOSITION 6.39. *Let $F : \mathbf{R}^n \to \mathbf{R}^m$ be a Lipschitz function, $x_0, h \in \mathbf{R}^n$. Then $v \in DF(x_0)(h)$ if and only if there exists a sequence $\{t_n\} \searrow 0$ such that*

$$v = \lim_n \frac{F(x_0 + t_n h) - F(x_0)}{t_n}.$$

In particular, if the one-side directional derivative at x_0 for the direction h exists, we have

$$DF(x_0)(h) = \{F'(x_0, h)\}.$$

PROOF. The proof is an immediate consequence of Proposition 6.32 and the fact that

$$\left| \frac{F(x_0 + t_n h_n) - F(x_0 + t_n h)}{t_n} \right| \le K|h_n - h|,$$

where K is the Lipschitz constant of F.   □

Before finishing this section we will study the relation between the generalized jacobian and the graphical coderivative.

THEOREM 6.40. *Let $F : \mathbf{R}^n \to \mathbf{R}^m$ be a Lipschitz function, $x_0, h \in \mathbf{R}^n$, and $v \in \mathbf{R}^m$. Then*

$$co\big(D^* F(x_0)(v)\big) = \{Mv : M \in \overline{J}F(x_0)\}. \tag{6.5}$$

PROOF. It is enough to take convex hulls in the formula of Proposition 6.37, and invoke Proposition 6.26.   □

Let us observe that Formula (6.5) allows us to obtain sufficient conditions, in terms of graphical coderivative, in order to have a Lipschitz Inverse Theorem. Namely,

THEOREM 6.41. *Let $F : \mathbf{R}^n \to \mathbf{R}^n$ be a Lipschitz function, $x_0 \in \mathbf{R}^n$. Assume that*

$$0 \in co\big(D^* F(x_0)(v)\big) \Rightarrow v = 0.$$

Then there exist W and V neighborhoods of $F(x_0)$ and x_0 respectively, and a Lipschitz function $G : W \to V$ such that $F(G(y)) = y$ for every $y \in W$ and $G(F(x)) = x$ for every $x \in V$.

6.5 PROBLEMS

(1) Prove that a function $f : \mathbf{R}^n \to \mathbf{R}$ is K-Lipschitz if and only if

$$epi f = epi f + epi(K| \cdot |),$$

which is also equivalent to the following equality:

$$f(x) = \inf_{w \in \mathbf{R}^n} \big(f(x) + K|x - w|\big).$$

(2) Prove that the composition of a C^1 function with a convex function is C^1 provided that it is differentiable.

(3) Give an example of a Lipschitz function $f : \mathbf{R} \to \mathbf{R}$ with no one-side directional derivatives at 0.

(4) Prove that $f^\circ(x_0, v) = \limsup_{x \to x_0, t \uparrow 0} \frac{f(x+tv)-f(x)}{t}$ and consequently

$$f^\circ(x_0, v) = \limsup_{x \to x_0, t \to 0} \frac{f(x + tv) - f(x)}{t}.$$

(5) Prove that $f^0(x, v)$ is usc in both variables.

(6) Give an example of a Lipschitz function $f : \mathbf{R} \to \mathbf{R}$ satisfying $f^\circ(0, 1) = a$ and $f^\circ(0, -1) = b$ for arbitrary $a, b \in \mathbf{R}$.

(7) Give an example of two Lipschitz functions $f, g : \mathbf{R} \to \mathbf{R}$ for which

$$\overline{\nabla}(f + g)(0) \neq \overline{\nabla}f(0) + \overline{\nabla}g(0).$$

(8) Consider the function $f : \mathbf{R}^2 \to \mathbf{R}$ defined by $f(x, y) = |x| - |y|$:

    (a) Observe that it is 1-Lipschitz.

    (b) Prove that $\partial^- f(0, 0) = \partial^+ f(0, 0) = \emptyset$.

    (c) Calculate the subderivative function $df(0, 0)$.

    (d) Prove that $f^\circ((0, 0), (h, k)) = |h| + |k|$ for every $(h, k) \in \mathbf{R}^2$.

    (e) Calculate $\overline{\nabla}f(0, 0)$.

    (f) Calculate $\partial^- f(x, y)$ and $\partial^+ f(x, y)$ at every point $(x, y) \in \mathbf{R}^2$.

    (g) Calculate $\partial_L f(0, 0)$ and $\partial^L f(0, 0)$. Deduce that

$$\partial_L f(0, 0) \cup \partial^L f(0, 0) \neq \overline{\nabla}f(0, 0).$$

(9) Let $f, g : \mathbf{R}^n \to \mathbf{R}$ be two Lipschitz functions. Assume that $f \leq g$. Let $x_0 \in \mathbf{R}^n$ such that $f(x_0) = g(x_0)$. Prove that

$$\overline{\nabla}f(x_0) \cap \overline{\nabla}g(x_0) \neq \emptyset.$$

Prove that the result is no longer true if we replace the generalized gradient by the subdifferential (Frechet, proximal or limiting).

(10) Let $S : X \rightrightarrows Y$ be a set valued function between two Banach spaces. We say that S is outer semicontinuous at $x_0 \in X$ whenever for every $\varepsilon > 0$ there exists a $r > 0$ such that $S(x) \subset S(x_0) + \varepsilon B$ provided that $x \in B(x_0, r)$. This definition extends the continuity one for single valued functions. Study the outer semicontinuity of the following functions $S : \mathbf{R} \rightrightarrows \mathbf{R}$:

(a) $S(x) = [-1, 1]$ if $x \neq 0$, $S(0) = \{0\}$.
(b) $S(x) = \{0\}$ if $x \neq 0$, $S(0) = [-1, 1]$.

(11) We say that a set valued function between two Banach spaces $S :$ $X \rightrightarrows Y$ is inner semicontinuous at $x_0 \in X$ whenever for every $y_0 \in$ $S(x_0)$ and every $\{x_n\}$ converging to x_0 there exists a sequence $\{y_n\}$ converging to y_0, such that $y_n \in S(x_n)$. For single valued functions this is continuity. Study the inner semicontinuity for the functions in the preceding problem.

(12) Prove that the subdifferential is neither inner nor outer semicontinuous. Hint: Consider a non C^1 differentiable function.

(13) Give an example of a function such that its limiting subdifferential is not inner semicontinuous.

(14) Find a function f with no outer semicontinuous limiting subdifferential. Hint: it is enough that $\partial_L f(x) = \emptyset$ at one point.

(15) Prove that for a Lipschitz function $f : \mathbf{R}^n \to \mathbf{R}$, the limiting subdifferential is an outer semicontinuous set valued function.

(16) Consider the function $f : \mathbf{R} \to \mathbf{R}$ defined by $f(t) = -\sqrt{|x| + 1}$. Prove that it is Lipschitz and all its subgradients are different from zero, but it does not have an inverse near $t = 0$. Observe that $0 \in \overline{\nabla} f(0)$.

(17) Consider the function $f : \mathbf{R} \to \mathbf{R}$ defined by $f(t) = t + t^2 \sin \frac{1}{t}$ if $t \neq 0$ and $f(0) = 0$. Prove that it is locally Lipschitz near $t = 0$. Calculate its generalized gradient at $t = 0$. Does it have a local inverse near 0?

(18) Prove that for a differentiable function $F : \mathbf{R}^n \to \mathbf{R}^m$ the graphical derivative is the differential.

(19) Calculate the graphical derivative and coderivative at 0 of the following functions $f : \mathbf{R} \to \mathbf{R}$:

(a) $f(t) = t \sin \frac{1}{t}$ for $t \neq 0$, $f(0) = 0$.
(b) $f(t) = t^2 \sin \frac{1}{t}$ for $t \neq 0$, $f(0) = 0$.

(20) Prove that for a C^1 function the graphical coderivative is the differential, but the result is no longer true in absence of continuity of the differential.

(21) Prove that for a continuous function $F : \mathbf{R}^n \to \mathbf{R}^m$ the following equalities hold:

$$D(-F)(x_0)(h) = -DF(x_0)(h) \quad D^*(-F)(x_0)(h) = -D^* F(x_0)(h).$$

(22) Let $F, G : \mathbf{R}^n \to \mathbf{R}^m$ be two continuous functions, $x_0, h \in \mathbf{R}^n$. Prove that

$$D(F + G)(x_0)(h) \subset DF(x_0)(h) + DG(x_0)(h)$$

provided that one of the functions is Lipschitz.

(23) Calculate the graphical derivative $Df(0)$ for the function $f(x) = \sqrt{|x|}$.

CHAPTER 7

Applications

We present three applications in this final chapter. They are only a small part of the many fields where nonsmooth analysis plays an important role nowadays. We have selected two applications to differential equations and one to the problem of solving general equations with the associated problem of guaranteeing the existence of fixed points.

7.1 FLOW INVARIANT SETS

The aim of this section is to study the following problem: Let $S \subset \mathbf{R}^n$ be a closed set, and $\varphi : \mathbf{R}^n \to \mathbf{R}^n$ a Lipschitz function. We consider the following differential equation with initial condition

$$x'(t) = \varphi(x(t)) \quad x(0) = x_0 \quad (E)$$

Under what hypotheses can we assure that the trajectories $x(t), t > 0$, remain in S provided that $x_0 \in S$? This is the so called Flow Invariant Set Problem.

Before tackling this problem, we refresh some ideas of initial value problems. Remember that an Euler polygonal, x_π, associated to a given partition $\pi = \{0 = t_0 < t_1 < \cdots < t_N = M\}$, is an approximate solution of problem (E) defined on $[0, M]$, as follows: $x_\pi(0) = x_0$, while for $i = 1, \ldots, N$ and $t \in (t_{i-1}, t_i]$ we define

$$x_\pi(t) = x_\pi(t_{i-1}) + (t - t_{i-1})\varphi(x_\pi(t_{i-1}))$$

It is well known that under mild conditions, for instance φ being Lipschitz, Euler polygonals converge to a solution of (E) provided that the diameter $\delta(\pi)$ of the partition goes to 0.

We return to the flow invariant set problem. In order to solve it, we introduce a function $f : [0, +\infty) \to [0, +\infty)$ defined as $f(t) = d_S(x(t))$. We have that $f(0) = 0$ provided that $x_0 \in S$, hence $x(t) \in S$ for every $t > 0$ if and only if f is nonincreasing, consequently, we have to study the subgradients of the function $f(t) = d_S(x(t))$. We invoke the Chain Rule,

An Introduction to Nonsmooth Analysis. http://dx.doi.org/10.1016/B978-0-12-800731-0.00007-2

and we deduce from Proposition 5.9 that

$$\partial f(t) \supset \{\langle x'(t), \zeta \rangle : \zeta \in \partial d_S(x(t))\} = \{\langle \varphi(x(t)), \zeta \rangle : \zeta \in \partial d_S(x(t))\}$$

and from Theorem 4.18 that

$$\partial f(t) \supset \{\langle \varphi(x(t)), \zeta \rangle : \zeta \in \hat{N}_S(x(t)) \cap \overline{B}(0, 1)\} \qquad (7.1)$$

provided that $x(t) \in S$.

Once established the preliminary results, we can enunciate the main theorem.

THEOREM 7.1. *Let $\varphi : \mathbf{R}^n \to \mathbf{R}^n$ be a Lipschitz function. A closed set $S \subset \mathbf{R}^n$ is flow invariant with respect to a differential equation $x' = \varphi(x)$ if and only if $\varphi(x) \in co(T_S(x))$ for every $x \in S$.*

PROOF. Let us assume first that S is a flow invariant set. We consider a point $x_0 \in S$ and the trajectory $x(t)$ that starts at $x_0 = x(0)$. The function f defined by $f(t) = d_S(x(t))$ if $t \geq 0$ and 0 otherwise is nonincreasing and consequently $\partial f(0) \subset (-\infty, 0]$ by Proposition 5.22. Hence Formula (7.1) implies $\langle \varphi(x_0), \zeta \rangle \leq 0$ for every $\zeta \in \hat{N}_S(x_0) \cap \overline{B}(0, 1)$. The restriction $\zeta \in \overline{B}(0, 1)$ can be trivially avoided. In order to finish, we only have to prove that $\langle v, \zeta \rangle \leq 0$ for every $\zeta \in \hat{N}_S(x_0)$, given that it implies that $v \in co(T_S(x_0))$. But this is just an easy consequence of Minkowski's Separation Theorem and Proposition 4.15.

Conversely, a vector $v \in T_S(x)$ satisfies $\langle v, \zeta \rangle \leq 0$ for every $\zeta \in \hat{N}_S(x)$ by Proposition 4.15 again. Finally, if $v \in co(T_S(x))$, also we have that $\langle v, \zeta \rangle \leq 0$ since inequalities are stable under convex combinations. In other words: $\langle \varphi(x), \zeta \rangle \leq 0$ holds for every $\zeta \in \hat{N}_S(x)$.

We consider an arbitrary Euler polygonal, x_π, defined on an interval $[0, M]$. We denote the nodes $x_\pi(t_i)$ by x_i. For every $i = 0, \ldots, N$ we select a point $s_i \in S$ such that $d_S(x_i) = |x_i - s_i|$, that is $s_i \in proj_S(x_i)$. We have that $\langle \varphi(s_i), x_i - s_i \rangle \leq 0$ since $x_i - s_i \in \hat{N}_S(s_i)$.

We are now going to estimate the distance from Euler's polygonal nodes to S

$$d_S(x_1) \leq |x_1 - x_0| = (t_1 - t_0)|\varphi(x_0)| \leq Ct_1,$$

where C is an upper bound of $|\varphi|$ in a neighborhood of x_0. We have that

$$
\begin{aligned}
d_S^2(x_2) &\leq |x_2 - s_1|^2 = |x_2 - x_1|^2 + |x_1 - s_1|^2 + 2\langle x_2 - x_1, x_1 - s_1 \rangle \\
&= |x_2 - x_1|^2 + d_S(x_1)^2 + 2(t_2 - t_1)\langle \varphi(x_1), x_1 - s_1 \rangle \\
&= |x_2 - x_1|^2 + d_S(x_1)^2 + 2(t_2 - t_1)\langle \varphi(x_1) - \varphi(s_i), x_1 - s_1 \rangle \\
&\quad + 2(t_2 - t_1)\langle \varphi(s_i), x_1 - s_1 \rangle \\
&\leq |x_2 - x_1|^2 + d_S(x_1)^2 + 2(t_2 - t_1)\langle \varphi(x_1) - \varphi(s_i), x_1 - s_1 \rangle \\
&\leq C^2(t_2 - t_1)^2 + d_S^2(x_1) + 2(t_2 - t_1)\langle \varphi(x_1) - \varphi(s_1), x_1 - s_1 \rangle \\
&\leq C^2(t_2 - t_1)^2 + d_S^2(x_1) + 2(t_2 - t_1)K|x_1 - s_1|^2 \\
&= C^2(t_2 - t_1)^2 + d_S^2(x_1) + 2(t_2 - t_1)K d_S^2(x_1),
\end{aligned}
$$

where K is the Lipschitz constant of φ. Analogously we have

$$
\begin{aligned}
d_S^2(x_{i+1}) &\leq C^2(t_{i+1} - t_i)^2 + d_S^2(x_i) + 2K(t_{i+1} - t_i)d_S^2(x_i) \\
&\leq C^2(t_{i+1} - t_i)\delta(\pi) + d_S^2(x_i) + 2K\delta(\pi)d_S^2(x_i).
\end{aligned}
$$

Joining these inequalities for consecutive indices i we get

$$
\begin{aligned}
d_S^2(x_{i+1}) &\leq C^2(t_{i+1} - t_0)\delta(\pi) + 2K\delta(\pi)(d_S^2(x_i) + \cdots + d_S^2(x_1)) \\
&\leq C^2 M \delta(\pi) + 2K\delta(\pi)(d_S^2(x_i) + \cdots + d_S^2(x_1)) \\
&\leq C^2 M \delta(\pi) + 2K C^2 M \delta(\pi)^2 + 2(2K\delta(\pi))^2(d_S^2(x_{i-1}) + \cdots + d_S^2(x_1)) \\
&\leq \cdots \\
&\leq C^2 M \delta(\pi)\left(1 + 2K\delta(\pi) + 2(2K\delta(\pi))^2 + 3(2K\delta(\pi))^3 + \cdots \right. \\
&\quad \left. + i(2K\delta(\pi))^i\right) \\
&\leq C^2 M \delta(\pi) \sum_{j=1}^{\infty} j(2K\delta(\pi))^j.
\end{aligned}
$$

The last series is convergent provided that $2K\delta(\pi) < 1$, hence $d_S^2(x_{i+1})$ goes to 0 when $\delta(\pi) \to 0$.

As the polygonal converges uniformly on $[0, M]$ to the unique solution of the initial value problem (E), we have that the trajectory remains in S until $t = M$, but M is arbitrary, hence S is flow invariant. $\square$

It is clear that with minor arrangements in the proof, this theorem can be reformulated in the following way:

THEOREM 7.2. *Let $\varphi : \mathbf{R}^n \to \mathbf{R}^n$ be a Lipschitz function. A closed set $S \subset \mathbf{R}^n$ is flow invariant with respect to a differential equation $x' = \varphi(x)$ if and only if $\langle \varphi(x), \zeta \rangle \leq 0$ for every $x \in S$ and every $\zeta \in \hat{N}_S(x)$.*

These theorems point out that the flow invariant character of a set S depends only on the geometrical behavior of the function φ on the border set, ∂S, since $T_S(x) = \mathbf{R}^n$ for interior points. This fact is easily observed in the next corollary.

COROLLARY 7.3. *For a Lipschitz function $\varphi : \mathbf{R}^n \to \mathbf{R}^n$, the closed unit ball, $\overline{B} \subset \mathbf{R}^n$, is flow invariant with respect to $x' = \varphi(x)$ if and only if $\langle \varphi(x), x \rangle \leq 0$ for $|x| = 1$.*

We finish with the following well known corollary:

COROLLARY 7.4. *The closed unit ball, $\overline{B} \subset \mathbf{R}^n$, is flow invariant with respect to a linear system $x' = Ax$ if and only if $\mathrm{Re}\lambda \leq 0$ for every eigenvalue of A, λ.*

7.2 VISCOSITY SOLUTIONS

First order Hamilton-Jacobi equations, in the stationary case, are of the form
$$F(x, u(x), \nabla u(x)) = 0 \quad (E),$$
where $F : \mathbf{R}^n \times \mathbf{R} \times \mathbf{R}^n \to \mathbf{R}$. Let us now introduce the notion of viscosity solutions.

DEFINITION 7.5. An *usc* function $u : \mathbf{R}^n \to \mathbf{R}$ is a viscosity subsolution of (E) if $F(x, u(x), \zeta) \leq 0$ for every $x \in \mathbf{R}^n$ and $\zeta \in \partial^+ u(x)$.

Similarly

DEFINITION 7.6. A *lsc* function $u : \mathbf{R}^n \to \mathbf{R}$ is a viscosity supersolution of (E) if $F(x, u(x), \zeta) \geq 0$ for every $x \in \mathbf{R}^n$ and $\zeta \in \partial^- u(x)$.

Continuous functions that are both sub- and supersolutions are called **viscosity solutions**. Of course we may consider viscosity solutions in an open subset of $\mathbf{R}^n$ instead of in the whole $\mathbf{R}^n$.

It is clear that if a viscosity solution u is differentiable at a point x_0, then we have $F(x_0, u(x_0), \nabla u(x_0)) = 0$. Hence a differentiable function is a viscosity solution of (E) if and only if it is a classical solution. Moreover Lipschitz viscosity solutions are classical solutions a.e. The following example shows that some equations may have a viscosity solution but not a classical solution.

Example. Let us consider the equation $|\nabla u(x)| = 1$, $x \in B(0, 1)$, with the boundary condition $u(x) = 0$ if $|x| = 1$. It is clear that this problem has no classical solution since Rolle's Theorem implies that if a differentiable function vanishes on the unit sphere, then its gradient vanishes at a point of the open unit ball. However the continuous function $u(x) = 1 - |x|$ is a viscosity solution that vanishes at the boundary. Let us prove it.

The function which defines the equation is $F(x, u(x), \nabla u(x)) = |\nabla u(x)| - 1$. The function u is differentiable at every $x \neq 0$, with $\nabla u(x) = -\frac{x}{|x|}$. When $x = 0$ we have $\partial^- u(0) = \emptyset$, while $\partial^+ u(0) = \overline{B}(0, 1)$, hence u is a viscosity solution.

Have we finished the example? The equation is also represented by the function
$$G(x, u(x), \nabla u(x)) = 1 - |\nabla u(x)|,$$
but $u(x) = 1 - |x|$ is not a subsolution because $G(0, u(0), \zeta) > 0$ if $|\zeta| < 1$. In other words, equations $F = 0$ and $-F = 0$ are not the same! Nevertheless we have the following:

PROPOSITION 7.7. *A continuous function $u_0 : \mathbf{R}^n \to \mathbf{R}$ is a viscosity solution of $F(x, u, \nabla u) = 0$ if and only if $-u_0$ is a viscosity solution of $-F(x, -u, -\nabla u) = 0$.*

PROOF. u_0 is a subsolution of $F(x, u, \nabla u) = 0$ if and only if $F(x, u_0(x), \zeta) \leq 0$ for every $x \in \mathbf{R}^n$ and $\zeta \in \partial^+ u_0(x) = -\partial^-(-u_0)(x)$. This is equivalent to
$$-F(x, -(-u_0)(x), -(-\zeta)) \geq 0$$
for every $x \in \mathbf{R}^n$ and $-\zeta \in \partial^-(-u_0)(x)$. In other words: $-u_0$ is a supersolution of $-F(x, -u, -\nabla u) = 0$. Similarly, we may prove that u_0 is a supersolution of $F(x, u, \nabla u) = 0$ if and only $-u_0$ is a subsolution of $-F(x, -u, -\nabla u) = 0$. □

The next example illustrates the fact that limits of classical solutions are not classical solutions.

Example. Let us consider the following continuous function $\varphi_n : \mathbf{R} \to \mathbf{R}$

$$\varphi_n(x) = \begin{cases} 1 & \text{if } x \leq 1 - \frac{1}{2n}, \\ \sqrt{4n - 1 - 4nx} & \text{if } 1 - \frac{1}{2n} \leq x \leq 1 - \frac{1}{4n}, \\ 4nx - 4n + 1 & \text{if } 1 - \frac{1}{4n} \leq x \leq 1, \\ 1 & \text{if } x \geq 1. \end{cases}$$

The function

$$x_n(t) = \begin{cases} 1 - |t| & \text{if } |t| \geq \frac{1}{2n}, \\ 1 - \frac{1}{4n} - nt^2 & \text{if } |t| \leq \frac{1}{2n}. \end{cases}$$

is an exact solution of equation

$$|x'| - \varphi_n(x) = 0$$

with boundary condition $x(-1) = x(1) = 0$. The sequence $\{x_n\}$ converges to the function $x_0(t) = 1 - |t|$, which is a viscosity solution of equation $|x'| - 1 = 0$ with the same boundary condition, but is not an exact solution since it is not differentiable at $t = 0$. Finally, the sequence $\{\varphi_n\}$ converges to 1 uniformly on compacts subsets of $[0, 1)$. This fact is a consequence of the following general result.

THEOREM 7.8. *Let $F, F_n : \mathbf{R}^m \times \mathbf{R} \times \mathbf{R}^m \to \mathbf{R}$ be continuous functions. Assume that $\{F_n\}$ converges to F uniformly on compacts. Assume also that u_n is a viscosity solution of*

$$F_n(x, v, \nabla v) = 0$$

for every n. If $u = \lim_n u_n$ uniformly on compacts, then u is a viscosity solution of

$$F(x, v, \nabla v) = 0.$$

PROOF. Let $x_0 \in \mathbf{R}^m$, and $\zeta \in \partial^- u(x_0)$. We know that there is a C^1 function φ, such that $\varphi(x_0) = u(x_0)$, $\varphi < u$ otherwise, and $\zeta = \nabla\varphi(x_0)$. We claim that for every k there is a point $x_{n_k} \in B(x_0, \frac{1}{k})$, such that $u_{n_k} - \varphi$ has a local minimum at x_{n_k}. Hence $\zeta_{n_k} = \nabla\varphi(x_{n_k}) \in \partial^- u_{n_k}(x_{n_k})$, and consequently

$$F_{n_k}(x_{n_k}, u_{n_k}(x_{n_k}), \zeta_{n_k}) \geq 0.$$

By letting k go to ∞ we get

$$F(x_0, u(x_0), \zeta) \geq 0$$

since $\lim_k x_{n_k} = x_0$, $\{u_{n_k}(x_{n_k})\}$ converges to $u(x_0)$ because $\{u_n\}$ converges to u uniformly on compact sets, and $\lim_k \zeta_{n_k} = \zeta$ because φ is C^1. In other words, u is a supersolution. Similarly we could prove that u is a subsolution.

Let us prove the claim. For a given k, let

$$\varepsilon = \frac{1}{3} \min \left\{ u(x) - \varphi(x) : |x - x_0| = \frac{1}{k} \right\}.$$

If n_k is such that $|u(x) - u_{n_k}(x)| < \varepsilon$ for every $x \in B(x_0, \frac{1}{k})$ the claim holds since the minimum that $u_{n_k} - \varphi$ attains over $\overline{B}(x_0, \frac{1}{k})$, by continuity, lies in $B(x_0, \frac{1}{k})$. $\square$

The next theorem is important since it validates the "viscosity" character of these solutions.

THEOREM 7.9. *Let F_n, $F : \mathbf{R}^m \times \mathbf{R} \times \mathbf{R}^m \to \mathbf{R}$ be continuous functions satisfying that $\{F_n\}$ converges uniformly on compact sets to F. Suppose that u_n is a C^2 solution of*

$$-\frac{1}{n}\Delta u + F_n(x, u, \nabla u) = 0. \tag{7.2}$$

Let us assume that $\{u_n\}$ converges uniformly on compact subsets to a continuous function u. Then u is a viscosity solution of

$$F(x, u, \zeta) = 0. \tag{7.3}$$

PROOF. We will prove that u is a subsolution. Let us consider a C^2 function φ such that $u(x_0) = \varphi(x_0)$ and $u < \varphi$ otherwise. Using the same argument as in the preceding theorem we can consider a sequence $\{x_n\}$ converging to x_0 such that $u_n - \varphi$ has a local maximum at x_n. We have:

$$\nabla u_n(x_n) = \nabla \varphi(x_n) \quad \Delta u_n(x_n) \leq \Delta \varphi(x_n),$$

which implies

$$F_n(x_n, u_n(x_n), \nabla \varphi(x_n)) = F_n(x_n, u_n(x_n), \nabla u_n(x_n)) = \frac{1}{n}\Delta u_n(x_n)$$

$$\leq \frac{1}{n}\Delta \varphi(x_n)$$

since u_n is solution of (7.2). Letting $n \to \infty$ we obtain

$$F(x_0, u(x_0), \nabla \varphi(x_0)) \leq 0.$$

Finally, let us consider $\zeta \in \partial^+ u(x_0)$. There exists a C^1 function, ψ such that $\psi(x_0) = u(x_0)$, and $\psi(x) > u(x)$ otherwise, such that $\zeta = \nabla \psi(x_0)$. It is well known that there exists a sequence $\{\varphi_n\}$ of C^2 functions converging uniformly to ψ on compact subsets, and satisfying also that $\{\nabla \varphi_n\}$ converges to $\nabla \psi$ uniformly on compacts too. We have, one more time, that there is a sequence $\{y_k\}$ converging to x_0 such that $u - \varphi_{n_k}$ has a local maximum at y_k. We may assume without loss of generality that the maximum is global and strict, hence

$$F(y_k, u(y_k), \nabla \varphi_{n_k}(y_k)) \leq 0,$$

and letting $k \to \infty$ we have

$$F(x_0, u(x_0), \nabla \psi(x_0)) \leq 0.$$

Hence u is a subsolution of (7.3). $\square$

From now on we will restrict to a particular case of stationary Hamilton-Jacobi equations, namely those of the form

$$u(x) + H(\nabla u(x)) = f(x), \tag{7.4}$$

where $H : \mathbf{R}^n \to \mathbf{R}$. In other words, $F(x, u, \zeta) = H(\zeta) + u - f(x)$. For this particular case of Hamilton-Jacobi equations, we have the following theorem, that we present without proof.

THEOREM 7.10.   *Let $H : \mathbf{R}^n \to \mathbf{R}$ be a continuous function, $f, g :$ $\mathbf{R}^n \to \mathbf{R}$ uniformly continuous. If u and v are bounded viscosity solutions of $u + H(\nabla u) = f$ and $v + H(\nabla v) = g$ respectively, then*

$$\|u - v\|_\infty \leq \|f - g\|_\infty.$$

Here, $\| \ \|_\infty$ represents the usual sup norm, namely

$$\|f\|_\infty = \sup\{|f(x)| : x \in \mathbf{R}^n\}.$$

Particularizing $f = g$, we obtain the following corollary.

COROLLARY 7.11. *Let $H : \mathbf{R}^n \to \mathbf{R}$ be a continuous function, $f : \mathbf{R}^n \to \mathbf{R}$ a uniformly continuous function. There exists at most one bounded viscosity solution of $u(x) + H(\nabla u(x)) = f(x)$.*

With respect to the existence of solutions, we have

THEOREM 7.12. *Assume that $H : \mathbf{R}^n \to \mathbf{R}$ is continuous and $f : \mathbf{R}^n \to \mathbf{R}$ is uniformly continuous and bounded. Then there exists a unique viscosity solution of $u(x) + H(\nabla u(x)) = f(x)$.*

7.3 SOLVING EQUATIONS

In this section we will study the problem of giving sufficient conditions to ensure that the equation $F(x) = 0$ has a solution. For the sake of simplicity we will restrict to the particular case of $F : \mathbf{R}^n \to \mathbf{R}^n$, which moreover can be applied to find fixed points. We may deal with the more general situation $F : \mathbf{R}^n \to \mathbf{R}^m$ in a similar way. A more interesting case is to consider F as a set valued function, but this problem is out of the scope of this book. We start with a Lemma which is the cornerstone of this section.

LEMMA 7.13. *Let $G : \mathbf{R}^n \to \mathbf{R}^n$ be a continuous function, $z_0 \in \mathbf{R}^n$. We define the scalar function $\varphi(x) = |z_0 - G(x)|$. Assume that $x_0 \in \mathbf{R}^n$ satisfies $\varphi(x_0) > 0$. Then we have*

$$\langle \zeta, h \rangle \le \left\langle \frac{G(x_0) - z_0}{|G(x_0) - z_0|}, v \right\rangle \tag{7.5}$$

whenever $\zeta \in \partial \varphi(x_0)$, $h \neq 0$, and $v \in DG(x_0)(h)$.

PROOF. Let $v \in DG(x_0)(h)$, Proposition 6.32 allows us to consider sequences $t_n \downarrow 0$ and $\{h_n\}$ converging to h such that

$$v = \lim_n \frac{G(x_0 + t_n h_n) - G(x_0)}{t_n}. \tag{7.6}$$

On the other hand $\zeta \in \partial \varphi(x_0)$ implies

$$\liminf_{x \to x_0} \frac{\varphi(x) - \varphi(x_0) - \langle \zeta, x - x_0 \rangle}{|x - x_0|} \ge 0. \tag{7.7}$$

Particularizing the Formula (7.7) for $x = x_0 + t_n h_n$ we obtain

$$0 \leq \liminf_n \frac{\varphi(x_0 + t_n h_n) - \varphi(x_0) - \langle \zeta, t_n h_n \rangle}{t_n |h_n|}$$

$$= \liminf_n \left[\frac{\varphi(x_0 + t_n h_n) - \varphi(x_0)}{t_n |h_n|} - \frac{\langle \zeta, h_n \rangle}{|h_n|} \right]$$

$$= \liminf_n \frac{\varphi(x_0 + t_n h_n) - \varphi(x_0)}{t_n |h|} - \left\langle \zeta, \frac{h}{|h|} \right\rangle,$$

hence

$$\langle \zeta, h \rangle \leq \liminf_n \frac{\varphi(x_0 + t_n h_n) - \varphi(x_0)}{t_n}$$

$$= \liminf_n \frac{|G(x_0 + t_n h_n) - z_0| - |G(x_0) - z_0|}{t_n}$$

$$= \liminf_n \frac{1}{t_n} \left[\frac{|G(x_0 + t_n h_n) - z_0|^2 - |G(x_0) - z_0|^2}{2|G(x_0) - z_0|} \right]$$

$$= \liminf_n \frac{1}{t_n} \left[\frac{|G(x_0 + t_n h_n) - G(x_0)|^2 + 2\langle G(x_0 + t_n h_n) - G(x_0), G(x_0) - z_0 \rangle}{2|G(x_0) - z_0|} \right]$$

$$= \left\langle v, \frac{G(x_0) - z_0}{|G(x_0) - z_0|} \right\rangle,$$

which establishes Formula (7.5).   □

If G is differentiable at x_0, we have that φ is differentiable at that point too, since $\varphi(x_0) > 0$, hence Formula (7.5) reduces to

$$\langle \nabla \varphi(x_0), h \rangle \leq \left\langle \frac{G(x_0) - z_0}{|G(x_0) - z_0|}, DG(x_0)(h) \right\rangle,$$

in other words, Lemma 7.13 is a sort of generalized Chain Rule. If we consider the function $G : \mathbf{R} \to \mathbf{R}$ defined by $G(t) = |t| + 1$, $x_0 = z_0 = 0$, $h = 1, \zeta = -1$ and $v = 1$, we obtain an example proving that we cannot always expect an equality in the formula.

Now, we introduce a property which will play a central role in the results of this section. Recall that S denotes the unit sphere of $\mathbf{R}^n$ as usual.

DEFINITION 7.14. Let $D \subset \mathbf{R}^n$, we say that a continuous function $G : D \to \mathbf{R}^n$ satisfies the *Derivative Condition*, *DC* in short, for the positive constant A if for every $x_0 \in D$ and $e \in S$ there exist $h \in S$ and $v \in DG(x_0)(h)$ such that $\langle e, v \rangle \geq A$.

For differentiable functions we may characterize this property in the following way.

PROPOSITION 7.15. *A differentiable function $G : U \to \mathbf{R}^n$ satisfies the DC for a positive constant A if and only if*

$$\inf\{|DG(x)(h)| : x \in U, h \in S\} > 0.$$

In particular $DG(x)$ must be onto for every $x \in U$, and we may take A equal to that inf. *If in addition G is C^1 and $K \subset U$ is compact, requiring $DG(x)$ to be onto for every $x \in K$ is also sufficient for $G : K \to \mathbf{R}^n$ to satisfy the DC.*

PROOF. We observe first that $A = \inf\{|DG(x)(h)| : x \in U, h \in S\} > 0$ implies that $Ker\,DG(x) = \{0\}$, equivalently $DG(x)$ is onto, for every $x \in U$. Assuming again that A is positive, we have that for every $x \in U$

$$A \leq \inf\{|DG(x)(h)| : h \in S\}.$$

Now, given $e \in S$, we choose $h \in S$ such that $e = \lambda DG(x)(h)$ for a positive λ, which is possible since $DG(x)$ is onto, hence $\langle e, DG(x)(h)\rangle = |DG(x)(h)| \geq A$. In other words, we have proved that G satisfies DC for the positive constant A.

Conversely, it is immediate to see that if G satisfies the DC, then $DG(x)$ is onto for every x since otherwise we may take a unit vector, e, orthogonal to the image of $DG(x)$ and we would have $\langle e, DG(x)(h)\rangle = 0$ for every h. Assume now that $\inf\{|DG(x)(h)| : x \in U, h \in S\} = 0$, then we may select $x_0 \in U$ and $h_0 \in S$ such that $|DG(x_0)(h_0)| < A$. If we start with $e = \frac{DG(x_0)(h_0)}{|DG(x_0)(h_0)|}$, in order to get a contradiction, it is enough to prove that

$$\langle e, DG(x_0)(h)\rangle \leq \langle e, DG(x_0)(h_0)\rangle = |DG(x_0)(h_0)|$$

for every $h \in S$. Let us observe first that we may assume that $DG(x_0)(h_0)$ minimizes the distance to the origin over the smooth manifold $DG(x_0)(S)$, and consequently the tangent space to that manifold at $DG(x_0)(h_0)$, that we denote by L, is orthogonal to $DG(x_0)(h_0)$, and consequently to e, by Lagrange Multipliers. For an arbitrary $h \in S$, we write $DG(x_0)(h) = \lambda DG(x_0)(h_0) + v$ with $v \in L$, hence

$$\langle e, DG(x_0)(h)\rangle = \lambda\langle e, DG(x_0)(h_0)\rangle + \langle e, v\rangle = \lambda\langle e, DG(x_0)(h_0)\rangle.$$

It only remains to observe that $\lambda \leq 1$, but this is an easy consequence of the fact that $DG(x_0)(h_0) + L$ is a support of the convex set $DG(x_0)(B)$ at $DG(x_0)(h_0)$.

When G is C^1, the function $|DG(x)(h)|$, $(x, h) \in K \times S$, is continuous and positive since $DG(x)$ is always onto. Compactness of $K \times S$ gives us the result. $\qquad\square$

We present an example that proves that we cannot drop differential continuity in the previous proposition.

Example. Consider the function $f : \mathbf{R} \to \mathbf{R}$ defined by

$$f(x) = x + x^2 \sin\frac{1}{x} + \int_0^{|x|} 2t\left(1 - \sin\frac{1}{t}\right) dt$$

if $x \neq 0$ and $f(0) = 0$. It is differentiable with derivative

$$f'(x) = 1 + 2|x| - \cos\frac{1}{x}$$

for $x \neq 0$ and $f'(0) = 1$. Hence $f'(x) \neq 0$ for every x. However we have that $\inf\{|f'(x)| : x \in [0, 1]\} = 0$ since $\lim_n f'(\frac{1}{2n\pi}) = 0$.

Now we are ready to state the main theorem of this section.

THEOREM 7.16. *Let* $G, H : B(a, r) \to \mathbf{R}^n, a \in \mathbf{R}^n$. *We denote* $F = G + H$.

Assume that

(1) *G is continuous and satisfies the DC for a positive constant A.*
(2) *H is L-Lipschitz, with $L < A$.*
(3) *$|F(a)| < r(A - L)$.*

Then, there exists $x \in B(a, r)$ such that $F(x) = 0$.

PROOF. Firstly we observe that we may assume without loss of generality that $H = 0$ and consequently $L = 0$, since F satisfies the DC for the positive constant $A - L$. Let us see it. Given $x \in B(s, r)$ and $e \in S$, there exist $h \in S$ and $v_1 \in DG(x)(h)$ such that $\langle e, v_1 \rangle > A$ since G satisfies the DC condition for the constant A. From Proposition 6.32, we deduce that

$$v_1 = \lim_n \frac{G(x + t_n h_n) - G(x)}{t_n}$$

for suitable sequences $\{h_n\}$ and $\{t_n\}$. Now, we consider the sequence

$$\lim_n \frac{H(x + t_n h_n) - H(x)}{t_n},$$

which is bounded since H is Lipschitz, hence it has a subsequence, corresponding to subsequences $\{h_{n_k}\}$ and $\{t_{n_k}\}$, convergent to a vector v_2 that belongs to $DH(x)(h)$ by Proposition 6.32 again. We have

$$v_1 + v_2 = \lim_n \frac{G(x + t_{n_k} h_{n_k}) - G(x)}{t_{n_k}} + \lim_n \frac{H(x + t_{n_k} h_{n_k}) - H(x)}{t_{n_k}}$$

$$= \lim_n \frac{F(x + t_{n_k} h_{n_k}) - F(x)}{t_{n_k}}$$

hence $v = v_1 + v_2 \in DF(x)(h)$ by Proposition 6.32 one more time. It only remains to observe that

$$\langle e, v \rangle = \langle e, v_1 \rangle + \langle e, v_2 \rangle \geq A + \langle e, v_2 \rangle \geq A - L$$

since $|v_2| \leq L$ because H is L-Lipschitz.

Once established this preliminary reduction, we proceed to prove the Theorem.

We denote $f(x) = |F(x)| = |G(x)|$, we are going to assume that $f(x) > 0$ for every $x \in B(a, r)$, and we will arrive at a contradiction. Let $x_0 \in B(a, r)$, $\zeta \in \partial f(x_0)$. We proceed to estimate $|\zeta|$. Now, we particularize the DC for $e = -\frac{G(x_0)}{|G(x_0)|}$ and deduce that there exist $h \in S$ and $v \in DG(x_0)(h)$ such that

$$A \leq \left\langle -\frac{G(x_0)}{|G(x_0)|}, v \right\rangle$$

hence

$$\langle \zeta, h \rangle \leq \left\langle \frac{G(x_0)}{|G(x_0)|}, v \right\rangle \leq -A$$

by Lemma 7.13. We conclude that $|\zeta| \geq A$. Finally, we apply the Decreasing Principle, Theorem 5.34, and arrive at the contradiction

$$0 < \inf\{f(x) : x \in B(a, r)\} \leq f(a) - rA = |F(a)| - rA < 0. \qquad \square$$

Let us observe that the theorem not only guarantees the existence of a solution, but moreover it gives us information about the location of the solution.

From Theorem 7.16, we may deduce several fixed point results. We present here two of them:

THEOREM 7.17. *Let $G : B(a, r) \to \mathbf{R}^n$ satisfy the DC for $A > 1$. If $G(a) \in B(a, r(A - 1))$, then G has a fixed point $x_0 \in B(a, r)$*

PROOF. It is enough to apply Theorem 7.16 with $K(x) = -x$ which is 1-Lipschitz. $\square$

THEOREM 7.18. *Let $K : B(a, r) \to \mathbf{R}^n$ be a L-Lipschitz function with $L < 1$, that is: a contractive function. K has a fixed point $x_0 \in B(a, r)$, provided that $K(a) \in B(a, r(1 - L))$.*

PROOF. We apply Theorem 7.16 with $G(x) = -x$ which satisfy DC for $A = 1$. $\square$

An immediate consequence of this Theorem is the celebrated *Contractive Mapping Theorem*.

We finish this chapter with a Surjective Function Theorem.

THEOREM 7.19. *Let $G : B(a, r) \to \mathbf{R}^n$ be a continuous function satisfying the DC for a positive constant A, and $K : B(a, r) \to \mathbf{R}^n$ a L-Lipschitz function with $L < A$. Let $F = G + K$. Then there exists a function $\phi : B(F(a), r(A - L)) \to \mathbf{R}^n$ such that $F(\phi(y)) = y$ for every $y \in B(F(a), r(A - L))$.*

PROOF. Fix $y \in B(F(a), r(A - L))$, it is enough to apply Theorem 7.16 to G and $K - y$, and observe that $|G(a) + K(a) - y| < r(A - L)$. With this we know that there exists $x \in B(a, r)$ such that $G(x) + K(x) - y = 0$ or equivalently $F(x) = y$. And once $\phi(y) = x$ is defined, the work is done. $\square$

What can we say about the function ϕ? We may consider it as a set valued function, $\phi : B(F(a), r(A - L)) \rightrightarrows B(a, r)$, defined by

$$\phi(y) = \{x \in B(a, r) : F(x) = y\}.$$

Theorem 7.19 ensures that $\phi(y) \neq \emptyset$ for every $y \in B(F(a), r(A - L))$. However it is easy to prove more.

PROPOSITION 7.20. *Under the same assumptions as in Theorem 7.19, the set valued function* $\phi : B(F(a), \frac{r}{3}(A - L)) \rightrightarrows B(a, \frac{r}{3})$ *satisfies the following condition: For every* $y_1, y_2 \in B(F(a), \frac{r}{3}(A - L))$ *and* $x_1 \in \phi(y_1)$, *there exists* $x_2 \in \phi(y_2)$ *such that*

$$|x_1 - x_2| \le \frac{1}{A - L}|y_1 - y_2|. \tag{7.8}$$

PROOF. Let $0 < \varepsilon < \frac{r}{3} - |a - x_1|$. We apply Theorem 7.19 to F defined on

$$B\left(x_1, \frac{1}{A - L}|y_1 - y_2| + \varepsilon\right) \subset B\left(x_1, \frac{2}{3}r + \varepsilon\right) \subset B(a, r).$$

As $y_2 \in B(y_1, |y_2 - y_1| + (A - L)\varepsilon) = B(F(x_1), |y_2 - y_1| + (A - L)\varepsilon)$, we deduce that there exists $x_2 \in B(x_1, \frac{1}{A-L}|y_1 - y_2| + \varepsilon)$ such that $F(x_2) = y_2$. Letting $\varepsilon \downarrow 0$ we get the inequality (7.8). $\square$

The following corollary is immediate.

COROLLARY 7.21. *Under the same hypotheses of Theorem 7.19, if in addition F is one to one, we have that ϕ is single valued and $\left(\frac{1}{A-L}\right)$-Lipschitz.*

7.4 PROBLEMS

(1) Let $A = (a_{i,j})$ be a 2×2 matrix. What can we say about A if the unit square $[-1, 1]^2$ is flow invariant with respect to the linear system $x' = Ax$?

(2) Let $\varphi(x) = -x + \psi(x)$ with ψ satisfying the following condition: $|\psi(x)| \le |x|$. Prove that every ball $\overline{B}(0, r)$ is flow invariant with respect to $x' = \varphi(x)$.

(3) Consider the function $G : \mathbf{R} \to \mathbf{R}$ defined by $G(t) = -t^{\frac{1}{3}}$, and observe, taking $x_0 = 0$, that Formula (7.5) is no longer true if $h = 0$.

(4) Let $G : \mathbf{R}^n \to \mathbf{R}^n$ be a linear function, observe that G satisfies the DC for a positive constant A if and only if it is an isomorphism. Find the best constant A.

(5) Let $G : \mathbf{R}^n \to \mathbf{R}^n$ be a linear function, assume that 1 is not an eigenvalue of G. Let $K : \mathbf{R}^n \to \mathbf{R}^n$ be a Lipschitz function. Assume that there exists a positive constant L such that $G + \lambda K$ has a fixed point provided that $|\lambda| < L$. Estimate the constant L.

(6) We define $G, K : \mathbf{R}^n \to \mathbf{R}^n$, where G is an isometry and K is contractive. Prove that equation $G(x) + K(x) = 0$ has a solution.

BIBLIOGRAPHY

[1] F.H. Clarke, Y.S. Ledyaev, R.J. Stern, P.R. Wolenski, Nonsmooth analysis and control theory, Graduate Texts in Mathematics, vol. 178, Springer-Verlag, New York, 1998.

[2] L. Evans, Partial differential equations, Graduate studies in Mathematics, vol. 19, American Mathematical Society, Providence, RI, 2010.

[3] G.J.O. Jameson, Topology and Normed Spaces, Chapman and Hall, London Halsted Press, New York, 1974.

[4] B.S. Mordukovich, Variational analysis and generalized differentiation I, II, Grundlehren der mathematischen Wissenschaften, vol. 331, Springer-Verlag, Berlin, Heidelberg, 2006.

[5] J.P. Penot, Calculus without derivatives, Graduate Texts in Mathematics, vol. 266, Springer, New York, 2013.

[6] R.T. Rockafellar, R.J-B. Wets, Variational analysis, Grundlehren der mathematischen Wissenschaften, vol. 317, Springer-Verlag, Berlin, Heidelberg, New York, 1998.

[7] W. Schirotzek, Nonsmooth Analysis, Springer-Verlag, Berlin, Heidelberg, 2007.

[8] K.T. Smith, Primer of Modern Analysis, Springer-Verlag, New York, Berlin, 1983.

This is a selfcontained book, and therefore the short but selected bibliography that we present is aimed to three goals, namely: background references, where to find the omitted proofs, and further reading.

Smith's book is a good reference for several variables analysis, it also includes a short introduction on measure theory and topology. On the other hand for an introduction to functional analysis Jameson's book is very convenient.

The general proofs of most of the theorems that we present in the finite dimensional setting, may be found in the book of Clarke et al. It includes a proof of the nonsmooth Multidirectional Mean Value Theorem too. You may look for the proof of Hahn-Banach Theorem in Jameson's book, meanwhile the proof of Rademacher's Theorem appears in the book of Smith among others. Rockafellar and Wets book includes a proof of Caratheodory's Theorem. For Ekeland Variational Principle, Mordukovich book is a good reference. Finally, for an introduction to viscosity solutions, see Evans for instance.

In order to study in depth nonsmooth analysis, [4], [5], [6], and [7], are good references. In my opinion, the book of Rockafellar and Wets is superb for a finite dimensional study, moreover, it includes an exhaustive bibliography, and excellent historical notes.

An Introduction to Nonsmooth Analysis. http://dx.doi.org/10.1016/B978-0-12-800731-0.00014-x

Index

Printed and bound by CPI Group (UK) Ltd, Croydon, CR0 4YY

03/10/2024

01040421-0014